DRAINING FOR PROFIT,

AND

DRAINING FOR HEALTH.

BY

GEO. E. WARING, Jr.,
ENGINEER OF THE DRAINAGE OF THE CENTRAL PARK, NEW-YORK.

"EVERY REPORTED CASE OF FAILURE IN DRAINAGE WHICH WE HAVE INVESTI-
"GATED, HAS RESOLVED ITSELF INTO IGNORANCE, BLUNDERING, BAD MANAGEMENT,
"OR BAD EXECUTION."—*Gisborne.*

ILLUSTRATED.

NEW YORK:
ORANGE JUDD & COMPANY,
41 PARK ROW.

LOVEJOY & SON,
ELECTROTYPERS AND STEREOTYPERS.
15 Vandewater street N. Y.

PREFATORY NOTE.

In presenting this book to the public the writer desires to say that, having in view the great importance of thorough work in land draining, and believing it advisable to avoid every thing which might be construed into an approval of half-way measures, he has purposely taken the most radical view of the whole subject, and has endeavored to emphasize the necessity for the utmost thoroughness in all draining operations, from the first staking of the lines to the final filling-in of the ditches.

That it is sometimes necessary, because of limited means, or limited time, or for other good reasons, to drain partially or imperfectly, or with a view only to temporary results, is freely acknowledged. In these cases the occasion for less completeness in the work must determine the extent to which the directions herein laid down are to be disregarded; but it is believed that, even in such cases, the principles on which those directions are founded should be always borne in mind.

NEWPORT, R. I., 1867.

LIST OF ILLUSTRATIONS.

TABLE OF CONTENTS.

CHAPTER VI.

WHAT DRAINING COSTS.

CHAPTER VII.

WILL IT PAY?

CHAPTER VIII.

HOW TO MAKE DRAINING TILES.

CHAPTER IX.

THE RECLAIMING OF SALT MARSHES.

CHAPTER X.

MALARIAL DISEASES.

CHAPTER XI.

HOUSE DRAINAGE AND TOWN SEWERAGE IN THEIR RELATIONS TO THE PUBLIC HEALTH.

CHAPTER I.

LAND TO BE DRAINED AND THE REASONS WHY.

Land which requires draining hangs out a sign of its condition, more or less clear, according to its circumstances, but always unmistakable to the practiced eye. Sometimes it is the broad banner of standing water, or dark, wet streaks in plowed land, when all should be dry and of even color; sometimes only a fluttering rag of distress in curling corn, or wide-cracking clay, or feeble, spindling, shivering grain, which has survived a precarious winter, on the ice-stilts that have stretched its crown above a wet soil; sometimes the quarantine flag of rank growth and dank miasmatic fogs.

To recognize these indications is the first office of the drainer; the second, to remove the causes from which they arise.

If a rule could be adopted which would cover the varied circumstances of different soils, it would be somewhat as follows: All lands, of whatever texture or kind, in which *the spaces between the particles of soil* are filled with water, (whether from rain or from springs,) within less than four feet of the surface of the ground, except during and *immediately* after heavy rains, require draining.

Of course, the *particles* of the soil cannot be made dry, nor should they be; but, although they should be moist themselves, they should be surrounded with air, not with water. To illustrate this: suppose that water be poured into a barrel filled with chips of wood until it runs over at the top. The spaces between the chips will be filled with water, and the chips themselves will absorb enough to become thoroughly wet;—this represents the worst condition of a wet soil. If an opening be made at the bottom of the barrel, the water which fills the spaces between the chips will be drawn off, and its place will be taken by air, while the chips themselves will remain wet from the water which they hold by absorption. A drain at the bottom of a wet field draws away the water from the free spaces between its particles, and its place is taken by air, while the particles hold, by attraction, the moisture necessary to a healthy condition of the soil.

There are vast areas of land in this country which do not need draining. The whole range of sands, gravels, light loams and moulds allow water to pass freely through them, and are sufficiently drained by nature, *provided*, they are as open at the bottom as throughout the mass. A sieve filled with gravel will drain perfectly; a basin filled with the same gravel will not drain at all. More than this, a sieve filled with the stiffest clay, if not "puddled,"* will drain completely, and so will heavy clay soils on porous and well drained subsoils. Money expended in draining such lands as do not require the operation is, of course, wasted; and when there is doubt as to the requirement,

---

* *Puddling* is the kneading or rubbing of clay with water, a process by which it becomes almost impervious, retaining this property until thoroughly dried, when its close union is broken by the shrinking of its parts. Puddled clay remains impervious as long as it is saturated with water, and it does not entirely lose this quality until it has been pulverized in a dry state.

A small proportion of clay is sufficient to injure the porousness of the soil by puddling.—A clay subsoil is puddled by being plowed over when too wet, and the injury is of considerable duration. Rain water collected in hollows of stiff land, by the simple movement given it by the wind, so puddles the surface that it holds the water while the adjacent soil is dry and porous.

The term *puddling* will often be used in this work, and the reader will understand, from this explanation, the meaning with which it is employed.

sufficient tests should be made before the outlay for so costly work is encountered.

There is, on the other hand, much land which only by thorough-draining can be rendered profitable for cultivation, or healthful for residence, and very much more, described as "ordinarily dry land," which draining would greatly improve in both productive value and salubrity.

The Surface Indications of the necessity for draining are various. Those of actual swamps need no description; those of land in cultivation are more or less evident at different seasons, and require more or less care in their examination, according to the circumstances under which they are manifested.

If a plowed field show, over a part or the whole of its surface, a constant appearance of dampness, indicating that, as fast as water is dried out from its upper parts, more is forced up from below, so that after a rain it is much longer than other lands in assuming the light color of dry earth, it unmistakably needs draining.

A pit, sunk to the depth of three or four feet in the earth, may collect water at its bottom, shortly after a rain;—this is a sure sign of the need of draining.

All tests of the condition of land as to water,—such as trial pits, etc.,—should be made, when practicable, during the wet spring weather, or at a time when the springs and brooks are running full. If there be much water in the soil, even at such times, it needs draining.

If the water of heavy rains stands for some time on the surface, or if water collects in the furrow while plowing, draining is necessary to bring the land to its full fertility.

Other indications may be observed in dry weather;—wide cracks in the soil are caused by the drying of clays, which, by previous soaking, have been pasted together; the curling of corn often indicates that in its early growth it has been prevented, by a wet subsoil, from sending down its roots below the reach of the sun's heat, where it would find, even in the dryest weather, sufficient moisture for a healththy growth; any *severe* effect of drought, except on poor sands and gravels, may be presumed to result from the same cause; and a certain wiryness of grass, together with a mossy or mouldy appearance of the ground, also indicate excessive moisture during some period of growth. The effects of drought are, of course, sometimes manifested on soils which do not require draining,—such as those poor gravels, which, from sheer poverty, do not enable plants to form vigorous and penetrating roots; but any soil of ordinary richness, which contains a fair amount of clay, will withstand even a severe drought, without great injury to its crop, if it is thoroughly drained, and is kept loose at its surface.

Poor crops are, when the cultivation of the soil is reasonably good, caused either by inherent poverty of the land, or by too great moisture during the season of early growth. Which of these causes has operated in a particular case may be easily known. Manure will correct the difficulty in the former case, but in the latter there is no real remedy short of such a system of drainage as will thoroughly relieve the soil of its surplus water.

The Sources of the Water in the soil are various. Either it falls directly upon the land as rain; rises into it from underlying springs; or reaches it through, or over, adjacent land.

The *rain water* belongs to the field on which it falls, and it would be an advantage if it could all be made to pass down through the first three or four feet of the soil, and be removed from below. Every drop of it is freighted with fertilizing matters washed out from the air, and in its descent through the ground, these are given up for the use of plants; and it performs other important work among the vegetable and mineral parts of the soil.

The *spring water* does not belong to the field,—not a

drop of it,—and it ought not to be allowed to show itself within the reach of the roots of ordinary plants. It has fallen on other land, and, presumably, has there done its appointed work, and ought not to be allowed to convert our soil into a mere outlet passage for its removal.

The *ooze water*,—that which soaks out from adjoining land,—is subject to all the objections which hold against spring water, and should be rigidly excluded.

But the *surface water* which comes over the surface of higher ground in the vicinity, should be allowed every opportunity, which is consistent with good husbandry, to work its slow course over our soil,—not to run in such streams as will cut away the surface, nor in such quantities as to make the ground inconveniently wet, but to spread itself in beneficent irrigation, and to deposit the fertilizing matters which it contains, then to descend through a well-drained subsoil, to a free outlet.

From whatever source the water comes, it cannot remain stagnant in any soil without permanent injury to its fertility.

The Objection to too much Water in the Soil will be understood from the following explanation of the process of germination, (sprouting,) and growth. Other grave reasons why it is injurious will be treated in their proper order.

The first growth of the embryo plant, (in the seed,) is merely a change of form and position of the material which the seed itself contains. It requires none of the elements of the soil, and would, under the same conditions, take place as well in moist saw-dust as in the richest mold. The conditions required are, the exclusion of light; a certain degree of heat; and the presence of atmospheric air, and moisture. Any material which, without entirely excluding the air, will shade the seed from the light, yield the necessary amount of moisture, and allow the accumulation of the requisite heat, will favor the chemical changes which, under these circumstances, take place in the living seed. In proportion as the heat is reduced by the chilling effect of evaporation, and as atmospheric air is excluded, will the germination of the seed be retarded; and, in case of complete saturation for a long time, absolute decay will ensue, and the germ will die.

The accompanying illustrations, (Figures 1, 2 and 3,) from the "Minutes of Information" on Drainage, submitted by the General Board of Health to the British Parliament in 1852, represent the different conditions of the soil as to moisture, and the effect of these conditions on the germination of seeds. The figures are thus explained by Dr. Madden, from whose lecture they are taken:

"Soil, examined mechanically, is found to consist entirely "of particles of all shapes and sizes, from stones and peb- "bles down to the finest powder; and, on account of their "extreme irregularity of shape, they cannot lie so close to "one another as to prevent there being passages between "them, owing to which circumstance soil in the mass is "always more or less *porous*. If, however, we proceed to "examine one of the smallest particles of which soil is "made up, we shall find that even this is not always solid, "but is much more frequently porous, like soil in the mass. "A considerable proportion of this finely-divided part of "soil, *the impalpable matter*, as it is generally called, is "found, by the aid of the microscope, to consist of *broken* "*down vegetable tissue*, so that when a small portion of "the finest dust from a garden or field is placed under the "microscope, we have exhibited to us particles of every "variety of shape and structure, of which a certain part is "evidently of vegetable origin.

"In these figures I have given a very rude representation "of these particles; and I must beg you particularly to "remember that they are not meant to represent by any "means accurately what the microscope exhibits, but are

" only designed to serve as a plan by which to illustrate " the mechanical properties of the soil. On referring to " Fig. 1, we perceive that there are two distinct classes of " pores,—first, the large ones, which exist *between* the par- " ticles of soil, and second, the very minute ones, which " occur in the particles themselves; and you will at the " same time notice that, " whereas all the larger " pores,—those between the " particles of soil, — com- " municate most freely with " each other, so that they " form canals, the small " pores, however freely they " may communicate with " one another in the interior " of the particle in which " they occur, have no direct " connection with the pores of the surrounding particles. " Let us now, therefore, trace the effect of this arrangement.

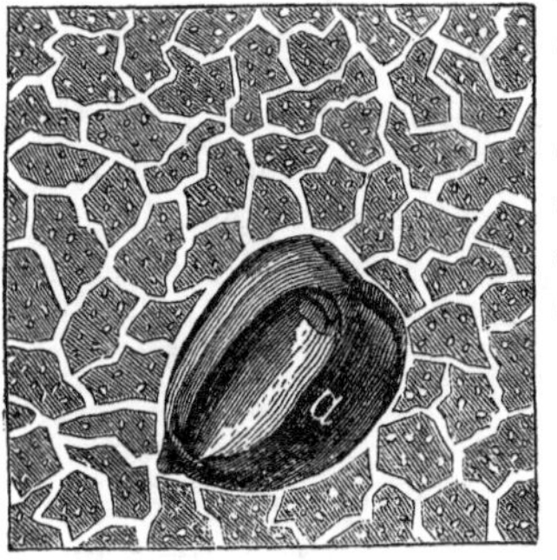

Fig. 1.—A DRY SOIL.

" In Fig. 1 we perceive that " these canals and pores are " all empty, the soil being " *perfectly dry;* and the " canals communicating free- " ly at the surface with the " surrounding atmosphere, " the whole will of course " be filled with air. If in " this condition a seed be " placed in the soil, at *a*, " you at once perceive that " it is freely supplied with air, *but there is no moisture;* " therefore, when soil is *perfectly dry*, a seed cannot grow.

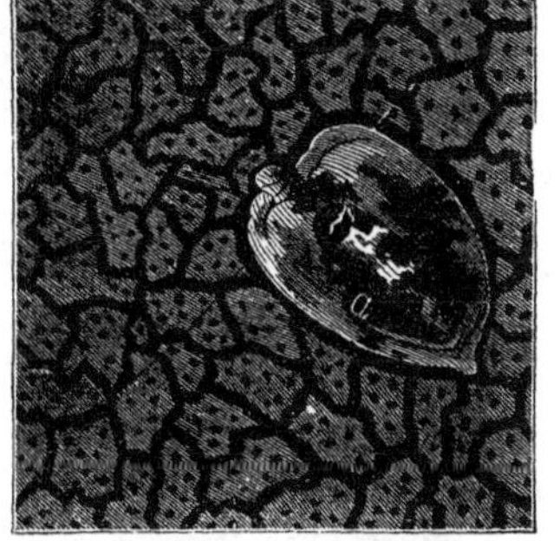

Fig. 2.—A WET SOIL.

" Let us turn our attention now to Fig. 2. Here we

"perceive that both the pores and canals are no longer "represented white, but black, this color being used to in- "dicate water; in this instance, therefore, water has taken "the place of air, or, in other words, the soil is *very wet.* "If we observe our seed *a* now, we find it abundantly "supplied with water, but *no air.* Here again, therefore, "germination cannot take place. If may be well to state "here that this can never occur *exactly* in nature, because, "water having the power of dissolving air to a certain "extent, the seed *a* in Fig. 2 is, in fact, supplied with a "*certain* amount of this necessary substance; and, owing "to this, germination does take place, although by no "means under such advantageous circumstances as it would "were the soil in a better condition.

"We pass on now to Fig. 3. Here we find a different "state of matters. The canals are open and freely sup- "plied with air, while the pores are filled with water; and, "consequently, you perceive "that, while the seed *a* has "quite enough of air from "the canals, it can never be "without moisture, as every "particle of soil which "touches it is well supplied "with this necessary in- "gredient. This, then, is "the proper condition of soil "for germination, and in "fact for every period of the "plant's development; and this condition occurs when the "soil is *moist*, but not *wet*,—that is to say, when it has the "color and appearance of being well watered, but when it "is still capable of being crumbled to pieces by the hands, "without any of its particles adhering together in the "familiar form of mud."

Fig. 3.—A DRAINED SOIL.

As plants grow under the same conditions, as to soil, that are necessary for the germination of seeds, the foregoing explanation of the relation of water to the particles of the soil is perfectly applicable to the whole period of vegetable growth. The soil, to the entire depth occupied by roots, which, with most cultivated plants is, in drained land, from two to four feet, or even more, should be main tained, as nearly as possible, in the condition represented in Fig. 3,—that is, the particles of soil should hold water by attraction, (absorption,) and the spaces between the particles should be filled with air. Soils which require drainage are not in this condition. When they are not saturated with water, they are generally dried into lumps and clods, which are almost as impenetrable by roots as so many stones. The moisture which these clods contain is not available to plants, and their surfaces are liable to be dried by the too free circulation of air among the wide fissures between them. It is also worthy of incidental remark, that the cracking of heavy soils, shrinking by drought, is attended by the tearing asunder of the smaller roots which may have penetrated them.

The Injurious Effects of Standing Water in the Subsoil may be best explained in connection with the description of a soil which needs under-draining. It would be tedious, and superfluous, to attempt to detail the various geological formations and conditions which make the soil unprofitably wet, and render draining necessary. Nor,—as this work is intended as a hand-book for practical use,—is it deemed advisable to introduce the geological charts and sections, which are so often employed to illustrate the various sources of under-ground water; interesting as they are to students of the theories of agriculture, and important as the study is, their consideration here would consume space, which it is desired to devote only to the reasons for, and the practice of, thorough-draining.

To one writing in advocacy of improvements, of any kind, there is always a temptation to throw a tub to the popular whale, and to suggest some make-shift, by which a certain advantage may be obtained at half-price. It is proposed in this essay to resist that temptation, and to adhere to the rule that "whatever is worth doing, is worth doing well," in the belief that this rule applies in no other department of industry with more force than in the draining of land, whether for agricultural or for sanitary improvement. Therefore, it will not be recommended that draining be ever confined to the wettest lands only; that, in the pursuance of a penny-wisdom, drains be constructed with stones, or brush, or boards; that the antiquated horse-shoe tiles be used, because they cost less money; or that it will, in any case, be economical to make only such drains as are necessary to remove the water of large springs. The doctrine herein advanced is, that, so far as draining is applied at all, it should be done in the most thorough and complete manner, and that it is better that, in commencing this improvement, a single field be really well drained, than that the whole farm be half drained.

Of course, there are some farms which suffer from too much water, which are not worth draining at present; many more which, at the present price of frontier lands, are only worth relieving of the water which stands on the surface; and not a few on which the quantity of stone to be removed suggests the propriety of making wide ditches, in which to hide them, (using the ditches, incidentally, as drains). A hand-book of draining is not needed by the owners of these farms; their operations are simple, and they require no especial instruction for their performance. This work is addressed especially to those who occupy lands of sufficient value, from their proximity to market, to make it cheaper to cultivate well, than to buy more land for the sake of getting a larger return from poor cultivation.

Wherever Indian corn is worth fifty cents a bushel, on the farm, it will pay to thoroughly drain every acre of land which needs draining. If, from want of capital, this cannot be done at once, it is best to first drain a portion of the farm, doing the work thoroughly well, and to apply the return from the improvement to its extension over other portions afterward.

In pursuance of the foregoing declaration of principles, it is left to the sagacity of the individual operator, to decide when the full effect desired can be obtained, on particular lands, without applying the regular system of depth and distance, which has been found sufficient for the worst cases. The directions of this book will be confined to the treatment of land which demands thorough work.

Such land is that which, at some time during the period of vegetation, contains stagnant water, at least in its subsoil, within the reach of the roots of ordinary crops; in which there is not a free outlet *at the bottom* for all the water which it receives from the heavens, from adjoining land, or from springs; and which is more or less in the condition of standing in a great, water-tight box, with openings to let water in, but with no means for its escape, except by evaporation at the surface; or, having larger inlets than outlets, and being at times "water-logged," at least in its lower parts. The subsoil, to a great extent, consists of clay or other compact material, which is not *impervious*, in the sense in which india-rubber is impervious, (else it could not have become wet,) but which is sufficiently so to prevent the free escape of water. The surface soil is of a lighter or more open character, in consequence of the cultivation which it has received, or of the decayed vegetable matter and the roots which it contains.

In such land the subsoil is wet,—almost constantly wet,—and the falling rain, finding only the surface soil in a condition to receive it, soon fills this, and often more than fills it, and stands on the surface. After the rain, come wind and sun, to dry off the standing water,—to dry out the free water in the surface soil, and to drink up the water of the subsoil, which is slowly drawn from below. If no spring, or ooze, keep up the supply, and if no more rain fall, the subsoil may be dried to a considerable depth, cracking and gaping open, in wide fissures, as the clay loses its water of absorption, and shrinks. After the surface soil has become sufficiently dry, the land may be plowed, seeds will germinate, and plants will grow. If there be not too much rain during the season, nor too little, the crop may be a fair one,—if the land be rich, a very good one. It is not impossible, nor even very uncommon, for such soils to produce largely, but they are always precarious. To the labor and expense of cultivation, which fairly earn a secure return, there is added the anxiety of chance; success is greatly dependent on the weather, and the weather may be bad. Heavy rains, after planting, may cause the seed to rot in the ground, or to germinate imperfectly; heavy rains during early growth may give an unnatural development, or a feeble character to the plants; later in the season, the want of sufficient rain may cause the crop to be parched by drought, for its roots, disliking the clammy subsoil below, will have extended within only a few inches of the surface, and are subject, almost, to the direct action of the sun's heat; in harvest time, bad weather may delay the gathering until the crop is greatly injured, and fall and spring work must often be put off because of wet.

The above is no fancy sketch. Every farmer who cultivates a retentive soil will confess, that all of these inconveniences conspire, in the same season, to lessen his returns, with very damaging frequency; and nothing is more common than for him to qualify his calculations with the proviso, "if I have a good season." He prepares his ground, plants his seed, cultivates the crop, "does his best,"—thinks he does his best, that is,—and trusts to Providence to send him good weather. Such farming is attended with too much uncertainty,—with too much *luck*,—to be satisfactory; yet, so long as the soil remains in its undrained condition, the element of luck will continue to play a very important part in its cultivation, and bad luck will often play sad havoc with the year's accounts.

Land of this character is usually kept in grass, as long as it will bring paying crops, and is, not unfrequently, only available for pasture; but, both for hay and for pasture, it is still subject to the drawback of the uncertainty of the seasons, and in the best seasons it produces far less than it might if well drained.

The effect of this condition of the soil on the health of animals living on it, and on the health of persons living near it, is extremely unfavorable; the discussion of this branch of the question, however, is postponed to a later chapter.

Thus far, there have been considered only the *effects* of the undue moisture in the soil. The manner in which these effects are produced will be examined, in connection with the manner in which draining overcomes them,—reducing to the lowest possible proportion, that uncertainty which always attaches to human enterprises, and which is falsely supposed to belong especially to the cultivation of the soil.

Why is it that the farmer believes, why should any one believe, in these modern days, when the advancement of science has so simplified the industrial processes of the world, and thrown its light into so many corners, that the word "mystery" is hardly to be applied to any operation of nature, save to that which depends on the always mysterious Principle of Life,—when the effect of any combination of physical circumstances may be foretold, with almost unerring certainty,—why should we believe that the success of farming must, after all, depend mainly on chance? That an intelligent man should submit the success of his own patient efforts to the operation of "luck;" that he should deliberately *bet* his capital, his toil, and his experience on having a good season, or a bad one,—this is not the least of the remaining mysteries. Some chance there must be in all things,—more in farming than in mechanics, no doubt; but it should be made to take the smallest possible place in our calculations, by a careful avoidance of every condition which may place our crops at the mercy of that most uncertain of all things—the weather; and especially should this be the case, when the very means for lessening the element of chance in our calculations are the best means for increasing our crops, even in the most favorable weather.

CHAPTER II.

HOW DRAINS ACT, AND HOW THEY AFFECT THE SOIL

For reasons which will appear, in the course of this work, the only sort of drain to which reference is here made is that which consists of a conduit of burned clay, (tile,) placed at a considerable depth in the subsoil, and enclosed in a compacted bed of the stiffest earth which can conveniently be found. Stone-drains, brush-drains, sod-drains, mole-plow tracks, and the various other devices for forming a conduit for the conveying away of the soakage-water of the land, are not without the support of such arguments as are based on the expediency of make-shifts, and are, perhaps, in rare cases, advisable to be used; but, for the purposes of permanent improvement, they are neither so good nor so economical as tile-drains. The arguments of this book have reference to the latter, (as the most perfect of all drains, thus far invented,) though they will apply, in a modified degree, to all underground conduits, so long as they remain free from obstructions. Concerning stone-drains, attention may properly be called to the fact that, (contrary to the general opinion of farmers,) they are very much more expensive than tile-drains. So great is the cost of cutting the ditches to the much greater size required for stone than for tiles, of handling the stones, of placing them properly in the ditches, and of covering them, after they are laid, with a suitable barrier to the rattling down of loose earth among them, that, as a mere question of first cost, it is far cheaper to buy tiles than to use stones, although these may lie on the surface of the field, and only require to be placed in the trenches. In addition to this, the great liability of stone-drains to become obstructed in a few years, and the certainty that tile-drains will, practically, last forever, are conclusive arguments in favor of the use of the latter. If the land is stony, it must be cleared; this is a proposition by itself, but if the sole object is to make drains, the best material should be used, and this material is not stone.

A well laid tile-drain has the following essential characteristics:—1. It has a free outlet for the discharge of all water which may run through it. 2. It has openings, at its joints, sufficient for the admission of all the water which may rise to the level of its floor. 3. Its floor is laid on a well regulated line of descent, so that its current may maintain a flow of uniform, or, at least, never decreasing rapidity, throughout its entire length.

Land which requires draining, is that which, at some time during the year, (either from an accumulation of the rains which fall upon it, from the lateral flow, or soakage, from adjoining land, from springs which open within it, or from a combination of two or all of these sources,) becomes filled with water, that does not readily find a natural outlet, but remains until removed by evaporation. Every considerable addition to its water wells up, and soaks its very surface; and that which is added after it is already brim full, must flow off over the surface, or lie in puddles upon it. Evaporation is a slow process, and it becomes more and more slow as the level of the water recedes from the surface, and is sheltered, by the overlying earth, from the action of sun and wind. Therefore, at least during the periods of spring and fall preparation of the land, during the early growth of plants, and often even in midsummer, the *water-table*,—the top of the water of saturation,—is within a few inches of the surface, preventing the natural descent of roots, and, by reason of the small space to receive fresh rains, causing an interruption of work for some days after each storm.

If such land is properly furnished with tile-drains, (having a clear and sufficient outfall, offering sufficient means of entrance to the water which reaches them, and carrying it, by a uniform or increasing descent, to the outlet,) its water will be removed to nearly, or quite, the level of the floor of the drains, and its water-table will be at the distance of some feet from the surface, leaving the spaces between the particles of all of the soil above it filled with air instead of water. The water below the drains stands at a level, like any other water that is dammed up. Rain water falling on the soil will descend by its own weight to this level, and the water will rise into the drains, as it would flow over a dam, until the proper level is again attained. Spring water entering from below, and water oozing from the adjoining land, will be removed in like manner, and the usual condition of the soil, above the water-table, will be that represented in Fig. 3, the condition which is best adapted to the growth of useful plants.

In the heaviest storms, some water will flow over the surface of even the dryest beach-sand; but, in a well drained soil the water of ordinary rains will be at once absorbed, will slowly descend toward the water-table, and will be removed by the drains, so rapidly, even in heavy clays, as to leave the ground fit for cultivation, and in a condition for steady growth, within a short time after the rain ceases.. It has been estimated that a drained soil has room between its particles for about one quarter of its bulk of water;—that is, four inches of drained soil contains free space enough to receive a rain-fall one inch in depth, and, by the same token, four feet of drained soil can receive twelve inches of rain,—more than is known to have ever fallen in twenty-four hours, since the deluge, and more than one quarter of the *annual* rain-fall in the United States.

As was stated in the previous chapter, the water which reaches the soil may be considered under two heads:

1st—That which reaches its surface, whether directly by rain, or by the surface flow of adjoining land.

2d—That which reaches it below the surface, by springs and by soakage from the lower portions of adjoining land.

The first of these is beneficial, because it contains fresh air, carbonic acid, ammonia, nitric acid, and heat, obtained from the atmosphere; and the flowage water contains, in addition, some of the finer or more soluble parts of the land over which it has passed. The second, is only so much dead water, which has already given up, to other soil, all that ours could absorb from it, and its effect is chilling and hurtful. This being the case, the only interest we can have in it, is to keep it down from the surface, and remove it as rapidly as possible.

The water of the first sort, on the other hand, should be arrested by every device within our reach. If the land is steep, the furrows in plowing should be run horizontally along the hill, to prevent the escape of the water over the surface, and to allow it to descend readily into the ground. Steep grass lands may have frequent, small, horizontal ditches for the same purpose. If the soil is at all heavy, it should not, when wet, be trampled by animals, lest it be puddled, and thus made less absorptive. If in cultivation, the surface should be kept loose and open, ready to receive all of the rain and irrigation water that reaches it.

In descending through the soil, this water, in summer, gives up heat which it received from the air and from the heated surface of the ground, and thus raises the temperature of the lower soil. The fertilizing matters which it has obtained from the air,—carbonic acid, ammonia and nitric acid,—are extracted from it, and held for the use of growing plants. Its fresh air, and the air which follows the descent of the water-table, carries oxygen to the organic and mineral parts of the soil, and hastens the rust and decay by which these are prepared for the uses of vegetation. The water itself supplies, by means of their power of absorption, the moisture which is needed by the particles of the soil; and, having performed its work, it goes down to the level of the water below, and, swelling the tide above the brink of the dam, sets the drains running, until it is all removed. In its descent through the ground, this water clears the passages through which it flows, leaving a better channel for the water of future rains, so that, in time, the heaviest clays, which will drain but imperfectly during the first one or two years, will pass water, to a depth of four or five feet, almost as readily as the lighter loams.

Now, imagine the drains to be closed up, leaving no outlet for the water, save at the surface. This amounts to a raising of the dam to that height, and additions to the water will bring the water-table even with the top of the soil. No provision being made for the removal of spring and soakage water, this causes serious inconvenience, and even the rain-fall, finding no room in the soil for its reception, can only lie upon, or flow over, the surface,—not yielding to the soil the fertilizing matters which it contains, but, on the contrary, washing away some of its finer and looser parts. The particles of the soil, instead of being furnished, by absorption, with a healthful amount of moisture, are made unduly wet; and the spaces between them, being filled with water, no air can enter, whereby the chemical processes by which the inert minerals, and the roots and manure, in the soil are prepared for the use of vegetation, are greatly retarded.

Instead of carrying the heat of the air, and of the surface of the ground, to the subsoil, the rain only adds so much to the amount of water to be evaporated, and increases, by so much, the chilling effect of evaporation.

Instead of opening the spaces of the soil for the more free passage of water and air, as is done by descending water, that which ascends by evaporation at the surface brings up soluble matters, which it leaves at the point where it becomes a vapor, forming a crust that prevents the free entrance of air at those times when the soil is dry enough to afford it space for circulation.

Instead of crumbling to the fine condition of a loam, as it does, when well drained, by the descent of water through it, heavy clay soil, being rapidly dried by evaporation, shrinks into hard masses, separated by wide cracks.

In short, in wet seasons, on such land, the crops will be greatly lessened, or entirely destroyed, and in dry seasons, cultivation will always be much more laborious, more hurried, and less complete, than if it were well drained.

The foregoing general statements, concerning the action of water in drained, and in undrained land, and of the effects of its removal, by gravitation, and by evaporation, are based on facts which have been developed by long practice, and on a rational application of well know principles of science. These facts and principles are worthy of examination, and they are set forth below, somewhat at length, especially with reference to *Absorption* and *Filtration; Evaporation; Temperature; Drought; Porosity* or *Mellowness;* and *Chemical Action.*

ABSORPTION AND FILTRATION.—The process of under-draining is a process of absorption and filtration, as distinguished from surface-flow and evaporation. The completeness with which the latter are prevented, and the former promoted, is the measure of the completeness of the improvement. If water lie on the surface of the ground until evaporated, or if it flow off over the surface, it will do harm; if it soak away through the soil, it will do good. The rapidity and ease with which it is absorbed, and, therefore, the extent to which under-draining is successful, depend on the physical condition of the soil, and on the manner in which its texture is affected by the drying action of sun and wind, and by the downward passage of water through it.

In drying, all soils, except pure sands, shrink, and occupy less space than when they are saturated with water. They shrink more or less, according to their composition, as will be seen by the following table of results obtained in the experiments of Schuebler:

1,000 Parts of	Will Contract Parts.	1,000 Parts of	Will Contract Parts.
Strong Limey Soil.......	50.	Pure Clay....	183.
Heavy Loam....	60.	Peat....................	200.
Brick Maker's Clay......	85.		

Professor Johnson estimates that peat and heavy clay shrink one-fifth of their bulk.

If soil be dried suddenly, from a condition of extreme wetness, it will be divived into large masses, or clods, separated by wide cracks. A subsequent wetting of the clods, which is not sufficient to expand it to its former condition, will not entirely obliterate the cracks, and the next drying will be followed by new fissures within the clods themselves; and a frequent repetition of this process will make the network of fissures finer and finer, until the whole mass of the soil is divided to a pulverulent condition. This is the process which follows the complete draining of such lands as contain large proportions of clay or of peat. It is retarded, in proportion to the amount of the free water in the soil which is evaporated from the surface, and in proportion to the trampling of the ground, when very wet. It is greatly facilitated by frost, and especially by deep frost.

The fissures which are formed by this process are, in time, occupied by the roots of plants, which remain and decay, when the crop has been removed, and which prevent the soil from ever again closing on itself so completely as before their penetration; and each season's crop adds new roots to make the separation more complete and more universal; but it is only after the water of saturation, which occupies the lower soil for so large a part of the year, has been removed by draining, that roots can penetrate to any considerable depth, and, in fact, the cracking of undrained soils, in drying, never extends beyond the separation into large masses, because each heavy rain, by saturating the soil and expanding it to its full capacity, entirely obliterates the cracks and forms a solid mass, in which the operation has to be commenced anew with the next drying.

Mr. Gisborne, in his capital essay on "Agricultural Drainage," which appeared in the *Quarterly Review*, No. CLXXI, says: "We really thought that no one was so ignorant as not to be aware that clay lands always shrink "and crack with drought, and the stiffer the clay the "greater the shrinking, as brickmakers well know. In the "great drought, 36 years ago, we saw in a very retentive "soil in the Vale of Belvoir, cracks which it was not "very pleasant to ride among. This very summer, on land "which, with reference to this very subject, the owner "stated to be impervious, we put a walking stick three "feet into a sun-crack, without finding a bottom, and the "whole surface was what Mr. Parkes, not inappropriately, "calls a network of cracks. When heavy rain comes "upon a soil in this state, of course the cracks fill, the clay "imbibes the water, expands, and the cracks are abolished. "But if there are four or five feet parallel drains in the "land, the water passes at once into them and is carried "off. In fact, when heavy rain falls upon clay lands in this "cracked state, it passes off too quickly, without adequate "filtration. Into the fissures of the undrained soil the roots "only penetrate to be perished by the cold and wet of the "succeeding winter; but in the drained soil the roots follow the threads of vegetable mold which have been "washed into the cracks, and get an abiding tenure. Earth "worms follow either the roots or the mold. Permanent "schisms are established in the clay, and its whole charac-"ter is changed. An old farmer in a midland county began "with 20-inch drains across the hill, and, without ever "reading a word, or, we believe, conversing with any one "on the subject, poked his way, step by step, to four or "five feet drains, in the line of steepest descent. Showing "us his drains this spring, he said: 'They do better year "by year; the water gets a habit of coming to them'—a very "correct statement of fact, though not a very philosophical "explanation."

Alderman Mechi, of Tiptree Hall, says: "Filtration "may be too sudden, as is well enough shown by our hot "sands and gravels; but I apprehend no one will ever "fear rendering strong clays too porous and manageable. "The object of draining is to impart to such soils the "mellowness and dark color of self drained, rich and fria-"ble soil. That perfect drainage and cultivation will do "this,is a well known fact. I know it in the case of my "own garden. How it does so I am not chemist enough "to explain in detail; but it is evident the effect is pro-"duced by the fibers of the growing crop intersecting "every particle of the soil, which they never could do be-"fore draining; these, with their excretions, decompose on "removal of the crop, and are acted on by the alternating "air and water, which also decompose and change, in a "degree, the inorganic substances of the soil. Thereby "drained land, which was, before, impervious to air and "water, and consequently unavailable to air and roots, "to worms, or to vegetable or animal life, becomes, by "drainage, populated by both, and is a great chemical "laboratory, as our own atmosphere is subject to all the "changes produced by animated nature."

Experience proves that the descent of water through the soil renders it more porous, so that it is easier for the water falling afterward to pass down to the drains, but no very satisfactory reason for this has been presented, beyond that which is connected with the cracking of the soil. The fact is well stated in the following extract from a letter to the *Country Gentleman:*

"A simple experiment will convince any farmer that the "best means of permanently deepening and mellowing the "soil is by thorough drainage, to afford a ready exit for all "surplus moisture. Let him take in spring, while wet, a "quantity of his hardest soil,—such as it is almost impossi- "ble to plow in summer,—such as presents a baked and "brick-like character under the influence of drought,—and "place it in a box or barrel, open at the bottom, and fre- "quently during the season let him saturate it with water. "He will find it gradually becoming more and more porous "and friable,—holding water less and less perfectly as the "experiment proceeds, and in the end it will attain a state "best suited to the growth of plants from its deep and "mellow character."

It is equally a fact that the ascent of water in the soil, together with its evaporation at the surface, has the effect of making the soil impervious to rains, and of covering the land with a crust of hard, dry earth, which forms a barrier to the free entrance of air. So far as the formation of crust is concerned, it is doubtless due to the fact that the water in the soil holds in solution certain mineral matters, which it deposits at the point of evaporation, the collection of these finely divided matters serving to completely fill the spaces between the particles of soil at the surface,—pasting them together, as it were. How far below the surface this direct action extends, cannot be definitely determined; but the process being carried on for successive years, accumulating a quantity of these fine particles, each season, they are, by cultivation, and by the action of heavy showers falling at a time when the soil is more or less dry, distributed through a certain depth, and ordinarily, in all probability, are most largely deposited at the top of the subsoil. It is found in practice that the first foot in depth of retentive soils is more retentive than that which lies below. If this opinion as to the cause of this greater imperviousness is correct, it will be readily seen how water, descending to the drains, by carrying these soluble and finer parts downward and distributing them more equally through the whole, should render the soil more porous.

Another cause of the retention of water by the surface soil, often a very serious one, is the puddling which clayey lands undergo by working them, or feeding cattle upon them, when they are wet. This is always injurious. By draining, land is made fit for working much earlier in the spring, and is sooner ready for pasturing after a rain, but, no matter how thoroughly the draining has been done, if there is much clay in the soil, the effect of the improvement will be destroyed by plowing or trampling, while very wet; this impervious condition will be removed in time, of course, but, while it lasts, it places us as completely at the mercy of the weather as we were before a ditch was dug.

In connection with the use of the word *impervious*, it should be understood that it is not used in its strict sense, for no substance which can be wetted by water is really impervious, and the most retentive soil will become wet. Gisborne states the case clearly when he says: "Is your "subsoil moister after the rains of mid-winter, than it is "after the drought of mid-summer? If it is, it will drain."

The proportion of the rain-fall which will filtrate through the soil to the level of the drains, varies with the composition of the soil, and with the effect that the draining has had upon them.

In a very loose, gravelly, or sandy soil, which has a perfect outlet for water below, all but the heaviest falls of rain will sink at once, while on a heavy clay, no matter how well it is drained, the process of filtration will be much more slow, and if the land be steeply inclined, some of the water of ordinarily heavy rains must flow off over the surface, unless, by horizontal plowing, or catch drains on the surface, its flow be retarded until it has time to enter the soil.

The power of drained soils to hold water, by absorption, is very great. A cubic foot of very dry soil, of favorable character, has been estimated to absorb within its particles, —holding no free water, or water of drainage,—about one-half its bulk of water; if this is true, the amount required to *moisten* a dry soil, four feet deep, giving no excess to be drained away, would amount to a rain fall of from 20 to 30 inches in depth. If we consider, in addition to this, the amount of water drained away, we shall see that the soil has sufficient capacity for the reception of all the rain water that falls upon it.

In connection with the question of absorption and filtration, it is interesting to investigate the movements of water in the ground. The natural tendency of water, in the soil as well as out of it, is to descend perpendicularly toward the center of the earth If it meet a flat layer of gravel lying upon clay, and having a free outlet, it will follow the course of the gravel,—laterally,—and find the outlet; if it meet water which is dammed up in the soil, and which has an outlet at a certain elevation, as at the floor of a drain, it will raise the general level of the water, and force it out through the drain; if it meet water which has no outlet, it will raise its level until the soil is filled, or until it accumulates sufficient pressure, (head,) to force its way through the adjoining lands, or until it finds an outlet at the surface.

The first two cases named represent the condition which it is desirable to obtain, by either natural or artificial drainage; the third case is the only one which makes drainage necessary. It is a fixed rule that water, descending in the soil, will find the *lowest* outlet to which there exists a channel through which it can flow, and that if, after heavy rains, it rise too near the surface of the ground, the proper remedy is to tap it at a lower level, and thus remove the water table to the proper distance from the surface. This subject will be more fully treated in a future chapter, in considering the question of the depth, and the intervals, at which drains should be placed.

Evaporation.—By evaporation is meant the process by which a liquid assumes the form of a gas or vapor, or "dries up." Water, exposed to the air, is constantly undergoing this change. It is changed from the liquid form, and becomes a vapor in the air. Water in the form of vapor occupies nearly 2000 times the space that it filled as a liquid. As the vapor at the time of its formation is of the same temperature with the water, and, from its highly expanded condition, requires a great *amount* of heat to maintain it as vapor, it follows that a given quantity of water contains, in the vapory form, many times as much heat as in the liquid form. This heat is taken from surrounding substances,—from the ground and from the air,—which are thereby made much cooler. For instance, if a shower moisten the ground, on a hot summer day, the drying up of the water will cool both the ground and the air. If we place a wet cloth on the head, and hasten the evaporation of the water by fanning, we cool the head; if we wrap a wet napkin around a pitcher of water, and place it in a current of air, the water in the pitcher is made cooler, by giving up its heat to the evaporating water of the napkin; when we sprinkle water on the floor of a room, its evaporation cools the air of the room.

So great is the effect of evaporation, on the temperature of the soil, that Dr. Madden found that the soil of a drained field, in which most of the water was removed from below, was $6\frac{1}{2}$° Far. warmer than a similar soil undrained, from which the water had to be removed by evaporation. This difference of $6\frac{1}{2}$° is equal to a difference of elevation of 1,950 feet.

It has been found, by experiments made in England, that the average evaporation of water from wet soils is equal to a depth of *two inches per month*, from May to August, inclusive; in America it must be very much greater than this in the summer months, but this is surely enough for the purposes of illustration, as two inches of water, over an acre of land, would weigh about *two hundred tons.* The amount of heat required to evaporate this is immense, and a very large part of it is taken from the soil, which, thereby, becomes cooler, and less favorable for a rapid growth. It is usual to speak of heavy,.wet lands as being " cold," and it is now seen why they are so.

If none of the water which falls on a field is removed by drainage, (natural or artificial,) and if none runs off from the surface, the whole rain-fall of a year must be removed by evaporation, and the cooling of the soil will be proportionately great. The more completely we withdraw this water from the surface, and carry it off in under-ground drains, the more do we reduce the amount to be removed by evaporation. In land which is well drained, the amount evaporated, even in summer, will not be sufficient to so lower the temperature of the soil as to retard the growth of plants; the small amount dried out of the particles of the soil, (water of absorption,) will only keep it from being raised to too great a heat by the mid-summer sun.

An idea of the amount of heat lost to the soil, in the evaporation of water, may be formed from the fact that to evaporate, by artificial heat, the amount of water contained in a rain-fall of two inches on an acre, (200 tons,) would require over 20 tons of coal. Of course a considerable—probably by far the larger,—part of the heat taken up in the process of evaporation is furnished by the air; but the amount abstracted from the soil is great, and is in direct proportion to the amount of water removed by this process; hence, the more we remove by draining, the more heat we retain in the ground.

The season of growth is lengthened by draining, because, by avoiding the cooling effects of evaporation, germination is more rapid, and the young plant grows steadily from the start, instead of struggling against the retarding influence of a cold soil.

Temperature.—The temperature of the soil has great effect on the germination of seeds, the growth of plants, and the ripening of the crops.

Gisborne says: "The evaporation of 1 lb. of water "lowers the temperature of 100 lbs. of soil 10°,—that is "to say, that, if to 100 lbs. of soil, holding all the water "it can by attraction, but containing no water of drainage, is added 1 lb. of water which it has no means of "discharging, except by evaporation, it will, by the time "that it has so discharged it, be 60° colder than it would "have been, if it had the power of discharging this 1 lb. "by filtration; or, more practically, that, if rain, entering "in the proportion of 1 lb. to 100 lbs. into a retentive "soil, which is saturated with water of attraction, is discharged by evaporation, it lowers the temperature of "that soil 10°. If the soil has the means of discharging "that 1 lb. of water by filtration, no effect is produced beyond what is due to the relative temperatures of the "rain and of the soil."

It has been established by experiment that four times as much heat is required to evaporate a certain quantity of water, as to raise the same quantity from the freezing to the boiling point.

It is, probably, in consequence of this cooling effect of evaporation, that wet lands are warmest when shaded, because, under this condition, evaporation is less active. Such lands, in cloudy weather, form an unnatural growth, such as results in the "lodging" of grain crops, from the deficient strength of the straw which this growth produces.

In hot weather, the temperature of the lower soil is, of course, much lower than that of the air, and lower than that of the water of warm rains. If the soil is saturated with water, the water will, of course, be of an even temperature with the soil in which it lies, but if this be drained off, warm air will enter from above, and give its heat to the soil, while each rain, as it falls, will also carry its heat with it. Furthermore, the surface of the ground is sometimes excessively heated by the summer sun, and the heat thus contained is carried down to the lower soil by the descending water of rains, which thus cool the surface and warm the subsoil, both beneficial.

Mr. Josiah Parkes, one of the leading draining engineers of England, has made some experiments to test the extent to which draining affects the temperature of the soil. The results of his observations are thus stated by Gisborne: "Mr. Parkes gives the temperature on a "Lancashire flat moss, but they only commence 7 inches "below the surface, and do not extend to mid-summer. "At that period of the year the temperature, at 7 inches, "never exceeded 66°, and was generally from 10° to 15° "below the temperature of the air in the shade, at 4 feet "above the earth. Mr. Parkes' experiments were made "simultaneously, on a drained, and on an undrained por-"tion of the moss; and the result was, that, on a mean "of 35 observations, the drained soil at 7 inches in depth "was 10° warmer than the undrained, at the same depth. "The undrained soil never exceeded 47°, whereas, after a "thunder storm, the drained reached 66° at 7 inches, and "48° at 31 inches. Such were the effects, at an early "period of the year, on a black bog. They suggest some "idea of what they were, when, in July or August, thunder "rain at 60° or 70° falls on a surface heated to 130°, and "carries down with it, into the greedy fissures of the earth, "its augmented temperature. These advantages, porous "soils possess by nature, and retentive ones only acquire "them by drainage."

Drained land, being more open to atmospheric circulation, and having lost the water which prevented the temperature of its lower portions from being so readily affected by the temperature of the air as it is when dry, will freeze to a greater depth in winter and thaw out earlier in the spring. The deep freezing has the effect to greatly pulverize the lower soil, thus better fitting it for the support of vegetation; and the earlier thawing makes it earlier ready for spring work.

Drought.—At first thought, it is not unnatural to suppose that draining will increase the ill effect of too dry seasons, by removing water which might keep the soil moist. Experience has proven, however, that the result is exactly the opposite of this. Lands which suffer most from drought are most benefited by draining,—more in their greater ability to withstand drought than in any other particular.

The reasons for this action of draining become obvious, when its effects on the character of the soil are examined. There is always the same amount of water in, and about, the surface of the earth. In winter there is more in the soil than in summer, while in summer, that which has been dried out of the soil exists in the atmosphere in the form of a *vapor*. It is held in the vapory form by *heat*, which may be regarded as *braces* to keep it distended. When vapor comes in contact with substances sufficiently colder than itself, it gives up its heat,—thus losing its braces,—contracts, becomes liquid water, and is deposited as dew.

Many instrances of this operation are familiar to all.

For instance, a cold pitcher in the summer robs the vapor in the air of its heat, and causes it to be deposited on its own surface,—of course the water comes from the atmosphere, not through the wall of the pitcher; if we breathe on a knife blade, it condenses, in the same manner, the moisture of the breath, and becomes covered with a film of water; stone-houses are damp in summer, because the inner surface of their walls, being cooler than the atmosphere, causes its moisture to be deposited in the manner described;* nearly every night, in summer, the cold earth receives moisture from the atmosphere in the form of dew; a single large head of cabbage, which at night is very cold, often condenses water to the amount of a gill or more.

The same operation takes place in the soil. When the air is allowed to circulate among its lower and cooler, (because more shaded,) particles, they receive moisture by the same process of condensation. Therefore, when, by the aid of under-drains, the lower soil becomes sufficiently loose and open to allow a circulation of air, the deposit of atmospheric moisture will keep it supplied with water, at a point easily accessible to the roots of plants.

If we wish to satisfy ourselves that this is practically correct, we have only to prepare two boxes of finely pulverized soil,—one three or four inches deep,—and the other fifteen or twenty inches deep, and place them in the sun, at midday, in summer. The thinner soil will soon be completely dried, while the deeper one, though it may have been previously dried in an oven, will soon accumulate a

* By leaving a space between the wall and the plastering, this moisture is prevented from being an annoyance, and if the inclosed space is not open from top to bottom, so as to allow a circulation of air, but little vapor will come in contact with the wall, and but an inconsiderable amount will be deposited.

large amount of water on those particles which, being lower and better sheltered from the sun's heat than the particles of the thin soil, are made cooler.

We have seen that even the most retentive soil,—the stiffest clay,—is made porous by the repeated passage of water from the surface to the level of the drains, and that the ability to admit air, which plowing gives it, is maintained for a much longer time than if it were usually saturated with water which has no other means of escape than by evaporation at the surface. The power of dry soils to absorb moisture from the air may be seen by an examination of the following table of results obtained by Schuebler, who exposed 1,000 grains of dried soil of the various kinds named to the action of the air:

Kind of Soil.	Amount of Water Absorbed in 24 Hours.
Common Soil........................	22 grains.
Loamy Clay..........................	26 grains.
Garden Soil..........................	45 grains.
Brickmakers' Clay..................	30 grains.

The effect of draining in overcoming drought, by admitting atmospheric vapor will, of course, be very much increased if the land be thoroughly loosened by cultivation, and especially if the surface be kept in an open and mellow condition.

In addition to the moisture received from the air, as above described, water is, in a porous soil, drawn up from the wetter subsoil below, by the same attractive force which acts to wet the whole of a sponge of which only the lower part touches the water;—as a hard, dry, compact sponge will absorb water much less readily than one which is loose and open, so the hard clods, into which undrained clay is dried, drink up water much less freely than they will do after draining shall have made them more friable.

The source of this underground moisture is the "water table,"—the level of the soil below the influence of the drains,—and this should be so placed that, while its water will easily rise to a point occupied by the feeding roots of the crop, it should yield as little as possible for evaporation at the surface.

Another source of moisture, in summer, is the deposit of dew on the surface of the ground. The amount of this is very difficult to determine, and accurate American experiments on the subject are wanting. Of course the amount of dew is greater here than in England, where Dr. Dalton, a skillful examiner of atmospheric phenomena, estimates the annual deposit of dew to equal a depth of five inches, or about one-fifth of the rain-fall. Water thus deposited on the soil is absorbed more or less completely, in proportion to the porosity of the ground.

The extent to which plants will be affected by drought depends, other things being equal, on the depth to which they send their roots. If these lie near the surface, they will be parched by the heat of the sun. If they strike deeply into the damper subsoil, the sun will have less effect on the source from which they obtain their moisture. Nothing tends so much to deep rooting, as the thorough draining of the soil. If the *free* water be withdrawn to a considerable distance from the surface, plants,—even without the valuable aid of deep and subsoil plowing,—will send their roots to great depths. Writers on this subject cite many instances in which the roots of ordinary crops "not mere hairs, but strong fibres, as large as pack-thread," sink to the depth of 4, 6, and in some instances 12 or 14 feet. Certain it is that, in a healthy, well aerated soil, any of the plants ordinarily cultivated in the garden or field will send their roots far below the parched surface soil; but if the subsoil is wet, cold, and soggy, at the time when the young crop is laying out its plan of future action, it will perforce accommodate its roots to the limited space which the comparatively dry surface soil affords.

It is well known among those who attend the meetings of the Farmers' Club of the American Institute, in New York, that the farm of Professor Mapes, near Newark, N. J., which maintains its wonderful fertility, year after year, without reference to wet or dry weather, has been rendered almost absolutely indifferent to the severest drought, by a course of cultivation which has been rendered possible only by under-draining. The lawns of the Central Park, which are a marvel of freshness, when the lands about the Park are burned brown, owe their vigor mainly to the complete drainage of the soil. What is true of these thoroughly cultivated lands, it is practicable to attain on all soils, which, from their compact condition, are now almost denuded of vegetation in dry seasons.

Porosity or Mellowness.—An open and mellow condition of the soil is always favorable for the growth of plants. They require heat, fresh air and moisture, to enable them to take up the materials on which they live, and by which they grow. We have seen that the heat of retentive soils is almost directly proportionate to the completeness with which their free water is removed by underground draining, and that, by reason of the increased facility with which air and water circulate within them, their heat is more evenly distributed among all those parts of the soil which are occupied by roots. The word *moisture*, in this connection, is used in contradistinction to *wetness*, and implies a condition of freshness and dampness,—not at all of saturation. In a saturated, a soaking-wet soil, every space between the particles is filled with water to the entire exclusion of the atmosphere, and in such a soil only aquatic plants will grow. In a *dry* soil, on the other hand, when the earth is contracted into clods and baked, almost as in an oven,—one of the most important conditions for growth being wanting,—nothing can thrive, save those plants which ask of the earth only an anchoring place, and seek their nourishment from the air. Both air plants and water plants have their wisely assigned places in the economy of nature, and nature provides them with ample space for growth. Agriculture, however, is directed to the production of a class of plants very different from either of these,—to those which can only grow to their greatest perfection in a soil combining, not one or two only, but all three of the conditions named above. While they require heat, they cannot dispense with the moisture which too great heat removes; while they require moisture, they cannot abide the entire exclusion of air, nor the dissipation of heat which too much water causes. The interior part of the pellets of a well pulverized soil should contain all the water that they can hold by their own absorptive power, just as the finer walls of a damp sponge hold it; while the spaces between these pellets, like the pores of the sponge, should be filled with air.

In such a soil, roots can extend in any direction, and to considerable depth, without being parched with thirst, or drowned in stagnant water, and, other things being equal, plants will grow to their greatest possible size, and all their tissues will be of the best possible texture. On rich land, which is maintained in this condition of porosity and mellowness, agriculture will produce its best results, and will encounter the fewest possible chances of failure. Of course, there are not many such soils to be found, and such absolute balance between warmth and moisture in the soil cannot be maintained at all times, and under all circumstances, but the more nearly it is maintained, the more nearly perfect will be the results of cultivation.

Chemical Action in the Soil.—Plants receive certain of their constituents from the soil, through their roots. The raw materials from which these constituents are obtained are the minerals of the soil, the manures which are artificially applied, water, and certain substances which are taken from the air by the absorptive action of the soil, or are brought to it by rains, or by water flowing over the surface from other land.

The mineral matters, which constitute the ashes of plants, when burned, are not mere accidental impurities which happen to be carried into their roots in solution in the water which supplies the sap, although they vary in character and proportion with each change in the mineral composition of the soil. It is proven by chemical analysis, that the composition of the ashes, not only of different species of plants, but of different parts of the same plant, have distinctive characters,—some being rich in phosphates, and others in silex; some in potash, and others in lime,—and that these characters are in a measure the same, in the same plants or parts of plants, without especial reference to the soil on which they grow. The minerals which form the ashes of plants, constitute but a very small part of the soil, and they are very sparsely distributed throughout the mass; existing in the interior of its particles, as well as upon their surfaces. As roots cannot penetrate to the interior of pebbles and compact particles of earth, in search of the food which they require, but can only take that which is exposed on their surfaces, and, as the oxydizing effect of atmospheric air is useful in preparing the crude minerals for assimilation, as well as in decomposing the particles in which they are bound up,—a process which is allied to the *rusting* of metals,—the more freely atmospheric air is allowed, or induced, to circulate among the inner portions of the soil, the more readily are its fertilizing parts made available for the use of roots. By no other process, is air made to enter so deeply, nor to circulate so readily in the soil, as by under-draining, and the deep cultivation which under-draining facilitates.

Of the manures which are applied to the land, those of a mineral character are affected by draining, in the same manner as the minerals which are native to the soil;

while organic, or animal and vegetable, manures, (especially when applied, as is usual, in an incompletely fermented condition,) absolutely require fresh supplies of atmospheric air, to continue the decomposition which alone can prepare them for their proper effect on vegetation.

If kept saturated with water, so that the air is excluded, animal manures lie nearly inert, and vegetable matters decompose but incompletely,—yielding acids which are injurious to vegetation, and which would not be formed in the presence of a sufficient supply of air. An instance is cited by H. Wauer where sheep dung was preserved, for five years, by excessive moisture, which kept it from the air. If the soil be saturated with water in the spring, and, in summer, (by the compacting of its surface, which is caused by evaporation,) be closed against the entrance of air, manures will be but slowly decomposed, and will act but imperfectly on the crop,—if, on the other hand, a complete system of drainage be adopted, manures, (and the roots which have been left in the ground by the previous crop,) will be readily decomposed, and will exercise their full influence on the soil, and on the plants growing in it.

Again, manures are more or less effective, in proportion as they are more or less thoroughly mixed with the soil. In an undrained, retentive soil, it is not often possible to attain that perfect *tilth*, which is best suited for a proper admixture, and which is easily given after thorough draining.

The soil must be regarded as the laboratory in which nature, during the season of growth, is carrying on those hidden, but indispensable chemical separations, combinations, and re-combinations, by which the earth is made to bear its fruits, and to sustain its myriad life. The chief demand of this laboratory is for free ventilation. The raw material for the work is at hand,—as well in the wet soil as in the dry; but the door is sealed, the damper is closed, and only a stray whiff of air can, now and then, gain entrance,—only enough to commence an analysis, or a combination, which is choked off when half complete, leaving food for sorrel, but making none for grass. We must throw open door and window, draw away the water in which all is immersed, let in the air, with its all destroying, and, therefore, all re-creating oxygen, and leave the forces of nature's beneficent chemistry free play, deep down in the ground. Then may we hope for the full benefit of the fertilizing matters which our good soil contains, and for the full effect of the manures which we add.

With our land thoroughly improved, as has been described, we may carry on the operations of farming with as much certainty of success, and with as great immunity from the ill effects of unfavorable weather, as can be expected in any business, whose results depend on such a variety of circumstances. We shall have substituted certainty for chance, as far as it is in our power to do so, and shall have made farming an art, rather than a venture.

CHAPTER III.

HOW TO GO TO WORK TO LAY OUT A SYSTEM OF DRAINS.

How to lay out the drains; where to place the outlet; where to locate the main collecting lines; how to arrange the laterals which are to take the water from the soil and deliver it at the mains; how deep to go; at what intervals; what fall to give; and what sizes of tile to use,—these are all questions of great importance to one who is about to drain land.

On the proper adjustment of these points, depend the *economy* and *effectiveness* of the work. Time and attention given to them, before commencing actual operations, will prevent waste and avoid failure. Any person of ordinary intelligence may qualify himself to lay out under-drains and to superintend their construction,—but the knowledge which is required does not come by nature. Those who have not the time for the necessary study and practice to make a plan for draining their land, will find it economical to employ an engineer for the purpose. In this era of railroad building, there is hardly a county in America which has not a practical surveyor, who may easily qualify himself, by a study of the principles and directions herein set forth, to lay out an economical plan for draining any ordinary agricultural land, to stake the lines, and to determine the grade of the drains, and the sizes of tile with which they should be furnished.

On this subject Mr. Gisborne says: "If we should give "a stimulus to amateur draining, we shall do a great deal "of harm. We wish we could publish a list of the moneys "which have been squandered in the last 40 years in amateur "draining, either ineffectually or with very imperfect effi- "ciency. Our own namo would be inscribed in the list for a "very respectable sum. Every thoughtless squire supposes "that, with the aid of his ignorant bailiff, he can effect a per- "fect drainage of his estate; but there is a worse man behind "the squire and the bailiff,—the draining conjuror. * * "* * * * These fellows never go direct about their "work. If they attack a spring, they try to circumvent "it by some circuitous route. They never can learn that "nature shows you the weakest point, and that you should "assist her,—that *hit him straight in the eye* is as good a "maxim in draining as in pugilism. * * * * * * "If you wish to drain, we recommend you to take advice. "We have disposed of the quack, but there is a faculty, "not numerous but extending, and whose extension ap- "pears to us to be indispensable to the satisfactory "progress of improvements by draining,—a faculty of "draining engineers. If we wanted a profession for a lad "who showed any congenial talent, we would bring him "up to be a draining engineer." He then proceeds to speak of his own experience in the matter, and shows that, after more than thirty years of intelligent practice, he employed Mr. Josiah Parkes to lay out and superintend his work, and thus effected a saving, (after paying all professional charges,) of fully twelve per cent. on the cost of the draining, which was, at the same time, better executed than any that he had previously done.

It is probable that, in nearly all amateur draining, the unnecessary frequency of the lateral drains; the extravagant size of the pipes used; and the number of useless angles which result from an unskillful arrangement, would amount to an expense equal to ten times the cost of the proper superintendence, to say nothing of the imperfect manner in which the work is executed. A common impression seems to prevail, that if a 2-inch pipe is good, a 3-inch pipe must be better, and that, generally, if draining is worth doing at all, it is worth overdoing; while the great importance of having perfectly fitting connections is not readily perceived. The general result is, that most of the tile-draining in this country has been too expensive for economy, and too careless for lasting efficiency.

It is proposed to give, in this chapter, as complete a description of the preliminary engineering of draining as can be concentrated within a few pages, and a hope is entertained, that it will, at least, convey an idea of the importance of giving a full measure of thought and ingenuity to the maturing of the *plan*, before the execution of the work is commenced. "Farming upon paper" has never been held in high repute, but draining upon paper is less a subject for objection. With a good map of the farm, showing the comparative levels of outlet, hill, dale, and plain, and the sizes and boundaries of the different inclosures, a profitable winter may be passed,—with pencil and rubber,—in deciding on a plan which will do the required work with the least possible length of drain, and which will require the least possible extra deep cutting; and in so arranging the main drains as to require the smallest possible amount of the larger and more costly pipes; or, if only a part of the farm is to be drained during the coming season, in so arranging the work that it will dovetail nicely with future operations. A mistake in actual work is costly, and, (being buried under the ground,) is not easily detected, while errors in drawing upon paper are always obvious, and are remedied without cost.

For the purpose of illustrating the various processes connected with the laying out of a system of drainage, the mode of operating on a field of ten acres will be detailed, in connection with a series of diagrams showing the progress of the work.

A Map of the Land is first made, from a careful survey. This should be plotted to a scale of 50 or 100 feet to the inch,* and should exhibit the location of obstacles which may interfere with the regularity of the drains,—such as large trees, rocks, etc., and the existing swamps, water courses, springs, and open drains. (Fig. 4.)

The next step is to locate the contour lines of the land, or the lines of equal elevation,—also called the *horizontal lines*,—which serve to show the shape of the surface. To do this, stake off the field into squares of 50 feet, by first running a base line through the center of the greatest length of the field, marking it with stakes at intervals of 50 feet, then stake other lines, also at intervals of 50 feet, perpendicular to the base line, and then note the position of the stakes on the maps; next, by the aid of an engineer's level and staff, ascertain the height, (above an imaginary plain below the lowest part of the field,) of the surface of the ground at each stake, and note this elevation at its proper point on the map. This gives a plot like Fig. 5. The best instrument with which to take these levels, is the ordinary telescope-level used by railroad engineers, shown in Fig. 6, which has a telescope with cross hairs intersecting each other in the center of the line of sight, and a "bubble" placed exactly parallel to this line. The instrument, fixed on a tripod, and so adjusted that it will turn to any point of the compass without disturbing the position of the bubble, will, (as will its "line of sight,") revolve in a perfectly horizontal plane. It is so placed as to command a view of a considerable stretch of the field, and its height above the imaginary plane is measured, an attendant places next to one of the stakes a levelling rod, (Fig. 7,) which is divided into feet and

* The maps in this book are, for convenience, drawn to a scale of 160 feet to the inch.

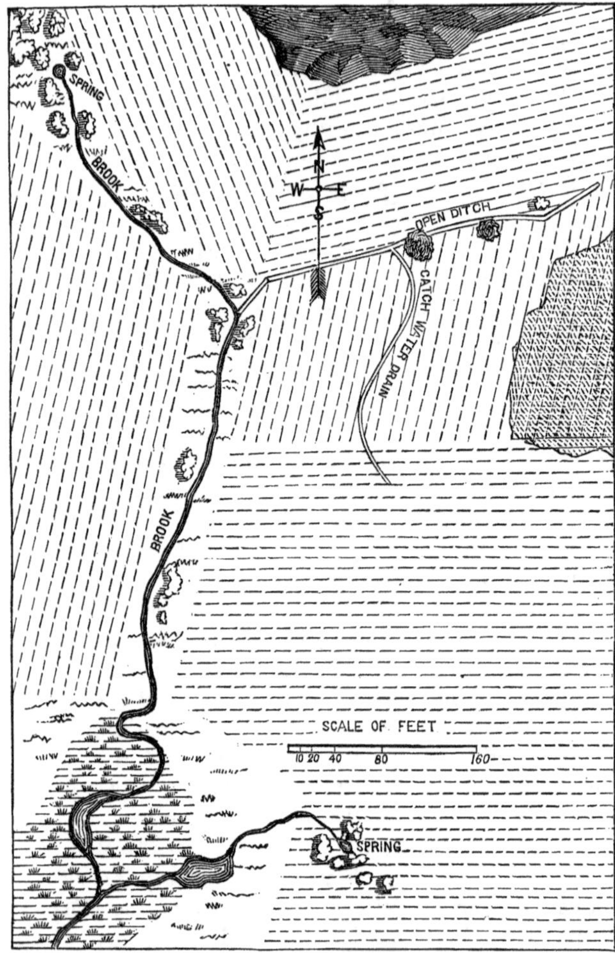

Fig. 4.—MAP OF LAND, WITH SWAMPS, ROCKS, SPRINGS, AND TREES. INTENDED TO REPRESENT A FIELD OF TEN ACRES BEFORE DRAINING.

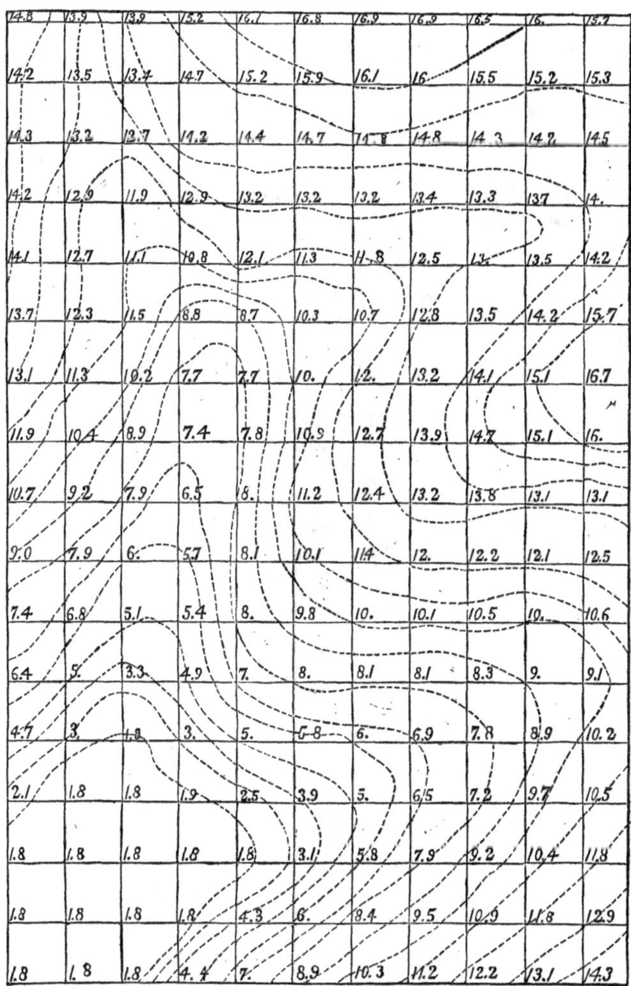

Fig. 5.—MAP WITH 50-FOOT SQUARES, AND CONTOUR LINES.

fractions of a foot, and is furnished with a movable target, so painted that its center point may be plainly seen. The attendant raises and lowers the target, until it comes exactly in the line of sight; its height on the rod denotes the height of the instrument above the level of the ground at that stake, and, as the height of the instrument

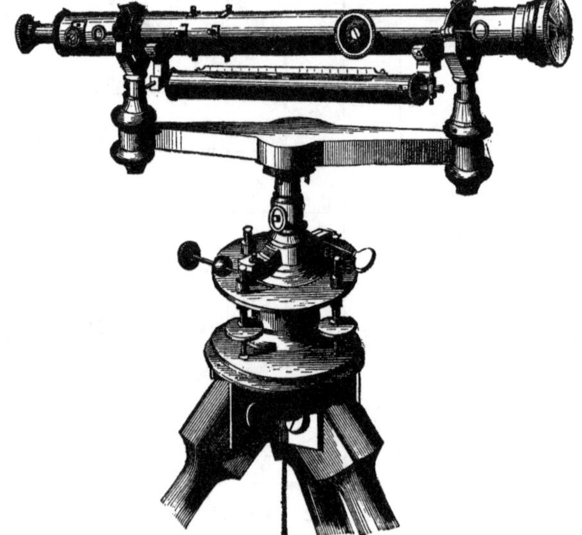

Fig. 6.—LEVELLING INSTRUMENT.*

above the imaginary plane has been reached, by subtracting one elevation from the other, the operator determines the height of the ground at that stake above the imaginary plane,—which is called the "*datum line.*"

The next operation is to trace, on the plan, lines following the same level, wherever the land is of the proper height for its surface to meet them. For the purpose of illustrating this operation, lines at intervals of elevation of

* The instrument from which this cut was taken, (as also Fig. 7,) was made by Messrs. Blunt & Nichols, Water-st., N. Y.

one foot are traced on the plan in Fig. 8. And these lines show, with sufficient accuracy for practical purposes, the elevation and rate of inclination of all parts of the field,—where it is level or nearly so, where its rise is rapid, and where slight. As the land rises one foot from the position of one line to the position of the line next above it, where the distance from one line to the next is great, the land is more nearly level, and when it is short the inclination is steeper. For instance, in the southwest corner of the plan, the land is nearly level to the 2-foot line; it rises slowly to the center of the field, and to the eastern side about one-fourth of the distance from the southern boundary, while an elevation coming down between these two valleys, and others skirting the west side of the former one and the southern side of the latter, are indicated by the greater nearness of the lines. The points at which the contour lines cross the section lines are found in the following manner: On the second line from the west side of the field we find the elevations of the 4th, 5th and 6th stakes from the southern boundary to be 1.9, 3.3, and 5.1. The contour lines, representing points of elevation of 2, 3, 4, and 5 feet above the *datum line*, will cross the 50-foot lines at their intersections, only where these intersections are marked in even feet. When they are marked with fractions of a foot, the lines must be made to cross at points between two intersections,—nearer to one or the other, according to their elevations,—thus between 1.9 and 3.3, the 2-foot and 3-foot contour lines must cross. The total difference of elevation, between the

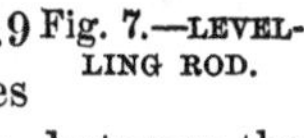

Fig. 7.—LEVELLING ROD.

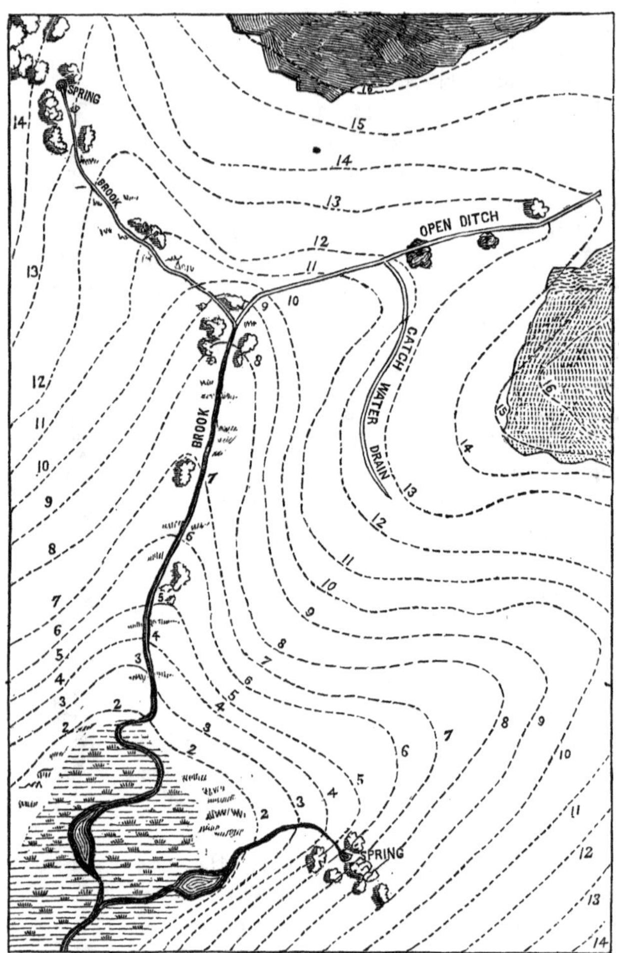

Fig. 8.—MAP WITH CONTOUR LINES.

two points is 3.3—1.9=1.4; $\frac{14}{14}$ of the space must be given to the even foot between the lines, and the 2-foot line should be $\frac{1}{14}$ of the space above the point 1.9;—the 3-foot line will then come $\frac{3}{14}$ below the point 3.3. In the same manner, the line from 3.3 to 5.1 is divided into 18 parts, of which 10 go to the space between the 4. and 5. lines, 7 are between 3.3 and the 4-foot line, and 1 between the 5-foot line and 5.1.

With these maps, made from observations taken in the field, we are prepared to lay down, on paper, our system of drainage, and to mature a plan which shall do the necessary work with the least expenditure of labor and material. The more thoroughly this plan is considered, the more economical and effective will be the work. Having already obtained the needed information, and having it all before us, we can determine exactly the location and size of each drain, and arrange, before hand, for a rapid and satisfactory execution of the work. The only thing that may interfere with the perfect application of the plan, is the presence of masses of underground rock, within the depth to which the drains are to be laid.* Where these are supposed to exist, soundings should be made, by driving a ¾-inch pointed iron rod to the rock, or to a depth of five feet where the rock falls away. By this means, measuring the distance from the soundings to the ranges of the stakes, we can denote on the map the shape and depth of sunken rocks. The shaded spot on the east side of the map, (Fig. 8,) indicates a rock three feet from the surface, which will be assumed to have been explored by sounding.

In most cases, it will be sufficient to have contour lines taken only at intervals of two feet, and, owing to the smallness of the scale on which these maps are engraved, and to avoid complication in the finished plan, where so

* The slight deviations caused by carrying the drains around large stones, which are found in cutting the ditches, do not affect the general arrangement of the lines.

much else must be shown, each alternate line is omitted. Of course, where drains are at once staked out on the land, by a practiced engineer, no contour lines are taken, as by the aid of the level and rod for the flatter portions, and by the eye alone for the steeper slopes, he will be able at once to strike the proper locations and directions; but for one of less experience, who desires to thoroughly mature his plan before commencing, they are indispensable; and their introduction here will enable the novice to understand, more clearly than would otherwise be possible, the principles on which the plan should be made.

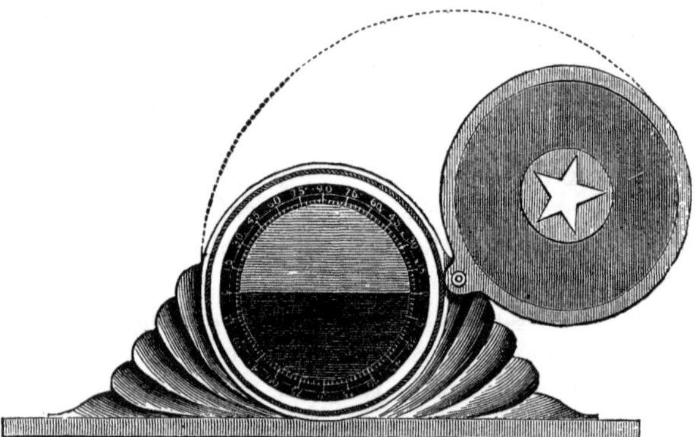

Fig. 9.—WELLS' CLINOMETER.

For preliminary examinations, and for all purposes in which great accuracy is not required, the little instrument shown in Fig. 9,—"Wells' Clinometer,"—is exceedingly simple and convenient. Its essential parts are a flat side, or base, on which it stands, and a hollow disk just half filled with some heavy liquid. The glass face of the disk is surrounded by a graduated scale that marks the angle at which the surface of the liquid stands, with reference to the flat base. The line 0.———0. being parallel to the base, when the liquid stands on that line, the flat side is horizontal; the line 90.———90. being perpendicular to the base, when the liquid stands on that line, the flat side is perpendicular or *plumb*. In like manner, the intervening angles are marked, and, by the aid of the following tables, the instrument indicates the rate of fall per hundred feet of horizontal measurement, and per hundred feet measured upon the sloping line.*

Table No. 1 shows the rise of the slope for 100 feet of the horizontal measurement. Example: If the horizontal distance is 100 feet, and the slope is at an angle of 15°, the rise will be $17\frac{633}{1000}$ feet.

Table No. 2 shows the rise of the slope for 100 feet of its own length. If the sloping line, (at an angle of 15°,) is 100 feet long, it rises 25.882 feet.

TABLE No. 1.				TABLE No. 2.			
DEG.	FEET.	DEG.	FEET.	DEG.	FEET.	DEG.	FEET.
5	8.749	50	119.175	5	8.716	50	76.604
10	17.633	55	142.815	10	17.365	55	81.915
15	26.795	60	173.205	15	25.882	60	86.602
20	36.397	65	214.451	20	34.202	65	90.631
25	46,631	70	274.748	25	42.262	70	93.969
30	57.735	75	373.205	30	50.—	75	96.593
35	70.021	80	567.128	35	57.358	80	98.481
40	83.910	85	1143.01	40	64.279	85	99.619
45	100.—			45	70.711		

With the maps before him, showing the surface features of the field, and the position of the under-ground rock, the drainer will have to consider the following points:

1. Where, and at what depth, shall the outlet be placed?

2. What shall be the location, the length and the depth of the main drain?

3. What subsidiary mains,—or collecting drains,—shall connect the minor valleys with the main?

4. What may best be done to collect the water of large springs and carry it away?

5. What provision is necessary to collect the water that flows over the surface of out-cropping rock, or

* The low price at which this instrument is sold, $1.50, places it within the reach of all.

along springy lines on side hills or under banks?

6. What should be the depth, the distance apart, the direction, and the rate of *fall*, of the lateral drains?

7. What kind and sizes of tile should be used to form the conduits?

8. What provision should be made to prevent the obstruction of the drains, by an accumulation of silt or sand, which may enter the tiles immediately after they are laid, and before the earth becomes compacted about them; and from the entrance of vermin?

1. The outlet should be at the lowest point of the boundary, unless, (for some especial reason which does not exist in the case under consideration, nor in any usual case,) it is necessary to seek some other than the natural outfall; and it should be deep enough to take the water of the main drain, and laid on a sufficient inclination for a free flow of the water. It should, where sufficient fall can be obtained without too great cost, deliver this water over a step of at least a few inches in height, so that the action of the drain may be seen, and so that it may not be liable to be clogged by the accumulation of silt, (or mud,) in the open ditch into which it flows.

2. The main drain should, usually, be run as nearly in the lowest part of the principal valley as is consistent with tolerable straightness. It is better to cut across the point of a hill, to the extent of increasing the depth for a few rods, than to go a long distance out of the direct course to keep in the valley, both because of the cost of the large tile used in the main, and of the loss of fall occasioned by the lengthening of the line. The main should be continued from the outlet to the point at which it is most convenient to collect the more remote sub-mains, which bring together the water of several sets of laterals. As is the case in the tract under consideration, the depth of the main is often restricted, in nearly level land, toward the upper end of the flat which lies next to the out-

let, by the necessity for a fall and the difficulty which often exists in securing a sufficiently low outlet. In such case, the only rule is to make it as deep as possible. When the fall is sufficient, it should be placed at such depth as will allow the laterals and sub-mains which discharge into it to enter at its top, and discharge above the level of the water which flows through it.

3. Subsidiary mains, or *sub-mains*, connecting with the main drains, should be run up the minor valleys of the land, skirting the bases of the hills. Where the valley is a flat one, with rising ground at each side, there should be a sub-main, to receive the laterals from *each* hill side. As a general rule, it may be stated, that the collecting drain at the foot of a slope should be placed on the line which is first reached by the water flowing directly down over its surface, before it commences its lateral movement down the valley; and it should, if possible, be so arranged that it shall have a uniform descent for its whole distance. The proper arrangement of these collecting drains requires more skill and experience than any other branch of the work, for on their disposition depends, in a great measure, the economy and success of the undertaking.

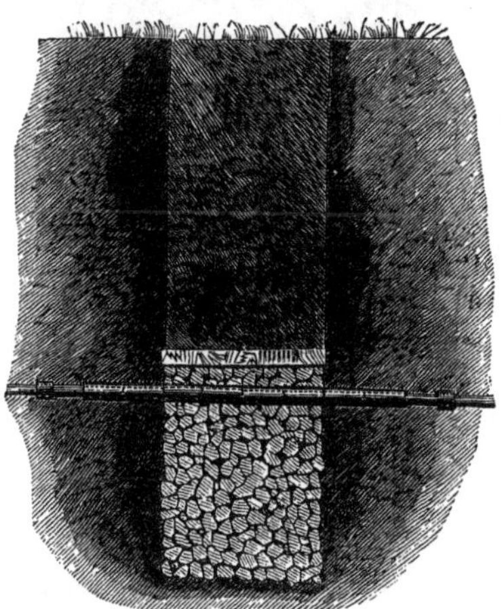

Fig. 10.—STONE PIT TO CONNECT SPRING WITH DRAIN.

4. Where springs exist, there should be some provision made for collecting their water in pits filled with loose

stone, gravel, brush or other rubbish, or furnished with several lengths of tile set on end, one above the other, or with a barrel or other vessel; and a line of tile of proper size should be run directly to a main, or sub-main drain. The manner of doing this by means of a pit filled with stone is shown in Fig. 10. The collection of spring water in a vertical tile basin is shown in Fig. 11.

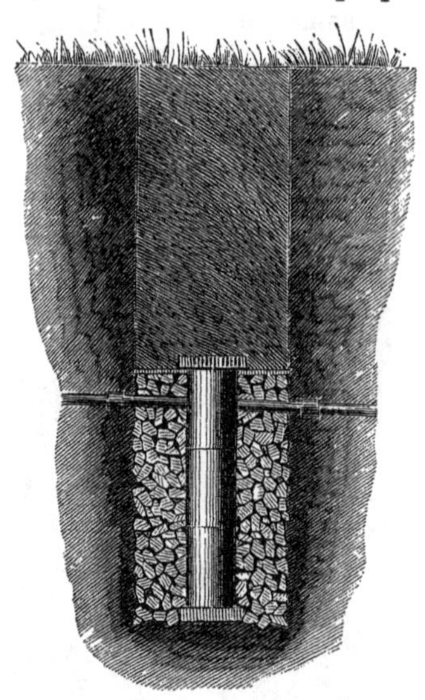

Fig. 11.—STONE AND TILE BASIN FOR SPRING WITH DRAIN.

5. Where a ledge of shelving rock, of considerable size, occurs on land to be drained, it is best to make some provision for collecting, at its base, the water flowing over its surface, and taking it at once into the drains, so that it may not make the land near it unduly wet. To effect this, a ditch should be dug along the base of the rock, and *quite down to it*, considerably deeper than the level of the proposed drainage; and this should be filled with small stones to that level, with a line of tile laid on top of the stones, a uniform bottom for the tile to rest upon being formed of cheap strips of board. The tile and stone should then be covered with inverted sods, with wood shavings, or with other suitable material, which will prevent the entrance of earth, (from the covering of the drain,) to choke them. The water, following down the surface of the rock, will rise through the stone work and, entering the tile, will flow off. This method may be used for springy hill sides.

6. The points previously considered relate only to the

collection of unusual quantities of water, (from springs and from rock surfaces,) and to the removal from the land of what is thus collected, and of that which flows from the minor or lateral drains.

The *lateral drains* themselves constitute the real drainage of the field, for, although main lines take water from the land on each side, their action in this regard is not usually considered, in determining either their depth or their location, and they play an exceedingly small part in the more simple form of drainage,—that in which a large tract of land, of perfectly uniform slope, is drained by parallel lines of equal length, all discharging into a single main, running across the foot of the field. The land would be equally well drained, if the parallel lines were continued to an open ditch beyond its boundary,—the main tile drain is only adopted for greater convenience and security. It will simplify the question if, in treating the *theory* of lateral drains, it be assumed that our field is of this uniform inclination, and admits of the use of long lines of parallel drains. In fact, it is best in practice to approximate as nearly as possible to this arrangement, because deviations from it, though always necessary in broken land, are always more expensive, and present more complicated engineering problems. If all the land to be drained had a uniform fall, in a single direction, there would be but little need of engineering skill, beyond that which is required to establish the depth, fall, and distance apart, at which the drains should be laid. It is chiefly when the land pitches in different directions, and with varying inclination, that only a person skilled in the arrangement of drains, or one who will give much consideration to the subject, can effect the greatest economy by avoiding unnecessary complication, and secure the greatest efficiency by adjusting the drains to the requirements of the land.

Assuming the land to have an unbroken inclination, so as to require only parallel drains, it becomes important to know how these parallel drains, (corresponding to the *lateral drains* of an irregular system,) should be made.

The history of land draining is a history of the gradual progress of an improvement, from the accomplishment of a single purpose, to the accomplishment of several purposes, and most of the instruction which modern agricultural writers have given concerning it, has shown too great dependence upon the teachings of their predecessors, who considered well the single object which they sought to attain, but who had no conception that draining was to be so generally valuable as it has become. The effort, (probably an unconscious one,) to make the theories of modern thorough-draining conform to those advanced by the early practitioners, seems to have diverted attention from some more recently developed principles, which are of much importance. For example, about a hundred years ago, Joseph Elkington, of Warwickshire, discovered that, where land is made too wet by under-ground springs, a skillful tapping of these,—drawing off their water through suitable conduits,—would greatly relieve the land, and for many years the Elkington System of drainage, being a great improvement on every thing theretofore practiced, naturally occupied the attention of the agricultural world, and the Board of Agriculture appointed a Mr. Johnstone to study the process, and write a treatise on the subject.

Catch-water drains, made so as to intercept a flow of surface water, have been in use from immemorial time, and are described by the earliest writers. Before the advent of the Draining Tile, covered drains were furnished with stones, boards, brush, weeds, and various other rubbish, and their good effect, very properly, claimed the attention of all improvers of wet land. When the tile first made its appearance in general practice, it was of what is called the "horse-shoe" form, and,—imperfect though it was,—it was better than anything that had preceded it, and was received with high approval, wherever it became known.

The general use of all these materials for making drains was confined to a system of *partial* drainage, until the publication of a pamphlet, in 1833, by Mr. Smith, of Deanston, who advocated the drainage of the whole field, without reference to springs. From this plan, but with important modifications in matters of detail, the modern system of tile draining has grown. Many able men have aided its progress, and have helped to disseminate a knowledge of its processes and its effects, yet there are few books on draining, even the most modern ones, which do not devote much attention to Elkington's discovery; to the various sorts of stone and brush drains; and to the manufacture and use of horse-shoe tile;—not treating them as matters of antiquarian interest, but repeating the instructions for their application, and allowing the reasoning on which their early use was based, to influence, often to a damaging extent, their general consideration of the modern practice of tile draining.

These processes are all of occasional use, even at this day, but they are based on no fixed rules, and are so much a matter of traditional knowledge, with all farmers, that instruction concerning them is not needed. The kind of draining which is now under consideration, has for its object the complete removal of all of the surplus water that reaches the soil, from whatever source, and the assimilation of all wet soils to a somewhat uniform condition, as to the ease with which water passes through them.

There are instances, as has been shown, where a large spring, overflowing a considerable area, or supplying the water of an annoying brook, ought to be directly connected with the under-ground drainage, and its flow neatly carried away; and, in other cases, the surface flow over large masses of rock should be given easy entrance into the tile; but, in all ordinary lands, whether swamps, springy hill sides, heavy clays, or light soils lying on retentive subsoil, all ground, in fact, which needs under-draining at all, should be laid dry above the level to which it is deemed best to place the drains;—not only secured against the wetting of springs and soakage water, but rapidly relieved of the water of heavy rains. The water table, in short, should be lowered to the proper depth, and, by permanent outlets at that depth, be prevented from ever rising, for any considerable time, to a higher level. This being accomplished, it is of no consequence to know whence the water comes, and Elkington's system need have no place in our calculations. As round pipes, with collars, are far superior to the "horse-shoe" tiles, and are equally easy to obtain, it is not necessary to consider the manner in which these latter should be used,—only to say that they ought not to be used at all.

The water which falls upon the surface is at once absorbed, settles through the ground, until it reaches a point where the soil is completely saturated, and raises the general water level. When this level reaches the floor of the drains, the water enters at the joints and is carried off. That which passes down through the land lying between the drains, bears down upon that which has already accumulated in the soil, and forces it to seek an outlet by rising into the drains.* For example, if a barrel, standing on end, be filled with earth which is saturated with water, and its bung be removed, the water of saturation, (that is, all which is not held by attraction *in* the particles of earth,) will be removed from so much of the mass as lies above the bottom of the bung-hole. If a bucket of water be now poured upon the top, it will not all run diagonally toward the opening; it will trickle down to the level of the water remaining in the barrel, and this level will rise and water will run off at the bottom of the orifice. In this manner, the water, even below the drainage level,

* Except from quite near to the drain, it is not probable that the water in the soil runs laterally towards it.

is changed with each addition at the surface. In a barrel filled with coarse pebbles, the water of saturation would maintain a nearly level surface; if the material were more compact and retentive, a true level would be attained only after a considerable time. Toward the end of the flow, the water would stand highest at the points furthest distant from the outlet. So, in the land, after a drenching rain, the water is first removed to the full depth, near the line of the drain, and that midway between two drains settles much more slowly, meeting more resistance from below, and, for a long time, will remain some inches higher than the floor of the drain. The usual condition of the soil, (except in very dry weather,) would be somewhat as represented in the accompanying cut, (Fig. 12.)

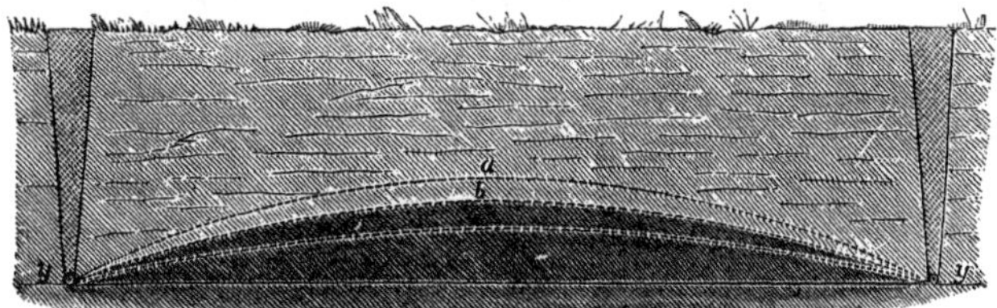

Fig. 12.—LINE OF SATURATION BETWEEN DRAINS.

Y Y are the drains. The curved line b is the line of saturation, which has descended from a, and is approaching c.

To provide for this deviation of the line of saturation, in practice, drains are placed deeper than would be necessary if the water sunk at once to the level of the drain floor, the depth of the drains being increased with the increasing distance between them.

Theoretically, every drop of water which falls on a field should sink straight down to the level of the drains, and force a drop of water below that level to rise into the drain and flow off. How exactly this is true in nature cannot be known, and is not material. Drains made in pursuance of this theory will be effective for any actual condition.

The *depth* to which the water table should be withdrawn depends, not at all on the character of the soil, but on the requirements of the crops which are to be grown upon it, and these requirements are the same in all soils,—consequently the depth should be the same in all.

What, then, shall that depth be? The usual practice of the most experienced drainers seems to have fixed four feet as about the proper depth, and the arguments against anything less than this, as well as some reasons for supposing that to be sufficient, are so clearly stated by Mr. Gisborne that it has been deemed best to quote his own words on the subject:

"Take a flower-pot a foot deep, filled with dry soil. "Place it in a saucer containing three inches of water. "The first effect will be, that the water will rise through "the hole in the bottom of the pot till the water which "fills the interstices between the soil is on a level with the "water in the saucer. This effect is by gravity. The "upper surface of this water is our water-table. From it "water will ascend by attraction through the whole "body of soil till moisture is apparent at the surface. Put "in your soil at 60°, a reasonable summer heat for nine "inches in depth, your water at 47°, the seven inches' "temperature of Mr. Parke's undrained bog; the attracted "water will ascend at 47°, and will diligently occupy "itself in attempting to reduce the 60° soil to its own "temperature. Moreover, no sooner will the soil hold "water of attraction, than evaporation will begin to carry "it off, and will produce the cold consequent thereon. "This evaporated water will be replaced by water of at- "traction at 47°, and this double cooling process will go "on till all the water in the water-table is exhausted. "Supply water to the saucer as fast as it disappears, and "then the process will be perpetual. The system of saucer- "watering is reprobated by every intelligent gardener; it "is found by experience to chill vegetation; besides which, "scarcely any cultivated plant can dip its roots into stag-"nant water with impunity. Exactly the process which "we have described in the flower-pot is constantly in "operation on an undrained retentive soil; the water-"table may not be within nine inches of the surface, but "in very many instances it is within a foot or eighteen "inches, at which level the cold surplus oozes into some "ditch or other superficial outlet. At eighteen inches, "attraction will, on the average of soils, act with consid-"erable power. Here, then, you have two obnoxious "principles at work, both producing cold, and the one "administering to the other. The obvious remedy is, to "destroy their *united* action; to break through their line "of communication. Remove your water of attraction "to such a depth that evaporation cannot act upon it, or "but feebly. What is that depth? In ascertaining this "point we are not altogether without data. No doubt "depth diminishes the power of evaporation rapidly. Still, "as water taken from a 30-inch drain is almost invariably "two or three degrees colder than water taken from four "feet, and as this latter is generally one or two degrees "colder than water from a contiguous well several feet "below, we can hardly avoid drawing the conclusion that "the cold of evaporation has considerable influence at 30 "inches, a much-diminished influence at four feet, and little "or none below that depth. If the water-table is removed "to the depth of four feet, when we have allowed 18 "inches of attraction, we shall still have 30 inches of de-"fence against evaporation; and we are inclined to be-"lieve that any prejudicial combined action of attraction "and evaporation is thereby well guarded against. The "facts stated seem to prove that less will not suffice.

"So much on the score of temperature; but this is not "all. Do the roots of esculents wish to penetrate into "the earth—at least, to the depth of some feet? We be-"lieve that they do. We are sure of the brassica tribe,

" of grass, and clover. All our experience and observation " deny the doctrine that roots only ramble when they are " stinted of food; that six inches well manured is quite " enough, better than more. Ask the Jerseyman; he " will show you a parsnip as thick as your thigh, and as " long as your leg, and will tell you of the advantages of " 14 feet of dry soil. You will hear of parsnips whose " roots descend to unsearchable depths. We will not " appeal to the Kentucky carrot, which was drawn out " by its roots at the antipodes; but Mr. Mechi's, if we " remember right, was a dozen feet or more. Three years " ago, in a midland county, a field of good land, in good " cultivation, and richly manured, produced a heavy crop " of cabbages. In November of that year we saw that " field broken into in several places, and at the depth of " four feet the soil (a tenacious marl, fully stiff enough for " brick-earth) was occupied by the roots of cabbage, not " sparingly—not mere capillæ—but fibres of the size of " small pack-thread. A farmer manures a field of four or " five inches of free soil reposing on a retentive clay, and " sows it with wheat. It comes up, and between the ker- " nel and the manure, it looks well for a time, but anon it " sickens. An Irish child looks well for five or six years, " but after that time potato-feeding, and filth, and hard- " ship, begin to tell. You ask what is amiss with the " wheat, and you are told that when its roots reach the " clay, they are poisoned. This field is then thorough- " drained, deep, at least four feet. It receives again from " the cultivator the previous treatment; the wheat comes " up well, maintains throughout a healthy aspect, and " gives a good return. What has become of the poison? " We have been told that the rain water filtered through " the soil has taken it into solution or suspension, and has " carried it off through the drains; and men who assume " to be of authority put forward this as one of the ad- " vantages of draining. If we believed it, we could not

" advocate draining. We really should not have the face " to tell our readers that water, passing through soils con- " taining elements prejudicial to vegetation, would carry " them off, but would leave those which are beneficial be- " hind. We cannot make our water so discriminating; the " general merit of water of deep drainage is, that it con- " tains very little. Its perfection would be that it should " contain nothing. We understand that experiments are " in progress which have ascertained that water, charged " with matters which are known to stimulate vegetation, " when filtered through four feet of retentive soil, comes " out pure. But to return to our wheat. In the first case, " it shrinks before the cold of evaporation and the cold of " water of attraction, and it sickens because its feet are " never dry; it suffers the usual maladies of cold and wet. " In the second case, the excess of cold by evaporation " is withdrawn; the cold water of attraction is removed " out of its way; the warm air from the surface, rushing " in to supply the place of the water which the drains re- " move, and the warm summer rains, bearing down with " them the temperature which they have acquired from " the upper soil, carry a genial heat to its lowest roots. " Health, vigorous growth, and early maturity are the " natural consequences. * * * * * * * * *

"The practice so derided and maligned referring to " deep draining has advanced with wonderful strides. " We remember the days of 15 inches; then a step to 20; a " stride to 30; and the last (and probably final) jump to 50, a " few inches under or over. We have dabbled in them all, " generally belonging to the deep section of the day. We " have used the words 'probably final,' because the first " advances were experimental, and, though they were jus- " tified by the results obtained, no one attempted to ex- " plain the principle on which benefit was derived from " them. The principles on which the now prevailing ' depth is founded, and which we believe to be true, go "far to show that we have attained all the advantages "which can be derived from the removal of water in "ordinary agriculture. We do not mean that, even in the "most retentive soil, water would not get into drains "which were laid somewhat deeper; but to this there "must be a not very distant limit, because pure clay, lying "below the depth at which wet and drought applied at "surface would expand and contract it, would certainly "part with its water very slowly. We find that, in coal "mines and in deep quarries, a stratum of clay of only a "few inches thick interposed between two strata of per-"vious stone will form an effectual bar to the passage of "water; whereas, if it lay within a few feet of the sur-"face, it would, in a season of heat and drought become "as pervious as a cullender. But when we have got rid "of the cold arising from the evaporation of free water, "have given a range of several feet to the roots of grass "and cereals, and have enabled retentive land to filter "through itself all the rain which falls upon its surface, "we are not, in our present state of knowledge, aware of "any advantage which would arise from further lowering "the surface of water in agricultural land. Smith, of "Deanston, first called prominent attention to the fertiliz-"ing effects of rain filtered through land, and to evils pro-"duced by allowing it to flow off the surface. Any one "will see how much more effectually this benefit will be "attained, and this evil avoided, by a 4-foot than a 2-foot "drainage. The latter can only prepare two feet of soil "for the reception and retention of rain, which two feet, "being saturated, will reject more, and the surplus must "run off the surface, carrying whatever it can find with it. "A 4-foot drainage will be constantly tending to have four "feet of soil ready for the reception of rain, and it will "take much more rain to saturate four feet than two. "Moreover, as a gimlet-hole bored four feet from the sur-"face of a barrel filled with water will discharge much "more in a given time than a similar hole bored at the "depth of two feet, so will a 4-foot drain discharge in a "given time much more water than a drain of two feet. "One is acted on by a 4-foot, and the other by a 2-foot "pressure."

If any single fact connected with tile-drainage is established, beyond all possible doubt, it is that in the stiffest clay soils ever cultivated, drains four feet deep will act effectually; the water will find its way to them, more and more freely and completely, as the drying of successive years, and the penetration and decay of the roots of successive crops, modify the character of the land, and they will eventually be practically so porous that,—so far as the ease of drainage is concerned,—no distinction need, in practice, be made between them and the less retentive loams. For a few years, the line of saturation between the drains, as shown in Fig. 11, may stand at all seasons considerably above the level of the bottom of the tile, but it will recede year by year, until it will be practically level, except immediately after rains.

Mr. Josiah Parkes recommends drains to be laid

"*At a minimum depth of four feet*, designed with the two-fold object of not only freeing the active soil from stagnant and injurious water, but of converting the water falling on the surface into an agent for fertilizing; no drainage being deemed efficient that did not both remove the water falling on the surface, and 'keep down the subterranean water at a depth exceeding the power of capillary attraction to elevate it near the surface.'"

Alderman Mechi says:

"Ask nineteen farmers out of twenty, who hold strong clay land, and they will tell you it is of no use placing deep four-foot drains in such soils —the water cannot get in; a horse's foot-hole (without an opening under it) will hold water like a basin; and so on. Well, five minutes after, you tell the same farmers you propose digging a cellar, well bricked, six or eight feet deep; what is their remark? 'Oh! it's of no use your making an underground cellar in our soil, you *can't keep the water* OUT!' Was there ever such an illustration of prejudice as this? What is a drain pipe but a small cellar full of air? Then, again, common sense tells us, you can't keep a light fluid under a heavy one. You might as well try to keep a cork under water, as to try and keep air under water. 'Oh! but then our soil is n't porous.' If not, how can it hold water so readily? I am led to these observations by the strong controversy I am having with some Essex folks, who protest that I am mad, or foolish, for placing 1-inch pipes, at four-foot depth, in strong clays. It is in vain I refer to the numerous proofs of my soundness, brought forward by Mr. Parkes, engineer to the Royal Agricultural Society, and confirmed by Mr. Pusey. They still dispute it. It is in vain I tell them *I cannot keep the rainwater out of* socketed pipes, twelve feet deep, that convey a spring to my farm yard. Let us try and convince this large class of doubters; for it is of *national* importance. Four feet of good porous clay would afford a far better meal to some strong bean, or other tap roots, than the usual six inches; and a saving of $4 to $5 per acre, in drainage, is no trifle.

"The shallow, or non-drainers, assume that tenacious subsoils are impervious or non-absorbent. This is entirely an erroneous assumption. If soils were impervious, how could they get wet?

"I assert, and pledge my agricultural reputation for the fact, that there are no earths or clays in this kingdom, be they ever so tenacious, that will not readily receive, filter, and transmit rain water to drains placed five or more feet deep.

"A neighbor of mine drained twenty inches deep in strong clay; the ground cracked widely; the contraction destroyed the tiles, and the rains washed the surface soils into the cracks and choked the drains. He has since abandoned shallow draining.

"When I first began draining, I allowed myself to be overruled by my obstinate man, Pearson, who insisted that, for top water, two feet was a sufficient depth in a veiny soil. I allowed him to try the experiment on two small fields; the result was, that nothing prospered; and I am redraining those fields at *one-half* the cost, five and six feet deep, at intervals of 70 and 80 feet.

"I found iron-sand rocks, strong clay, silt, iron, etc., and an enormous quantity of water, all *below* the 2-foot drains. This accounted at once for the sudden check the crops always met with in May, when they wanted to send their roots down, but could not, without going into stagnant water."

"There can be no doubt that it is the *depth* of the drain which regulates the escape of the surface water in a given time; regard being had, as respects extreme distances, to the nature of the soil, and a due capacity of the pipe. *The deeper the drain, even in the strongest soils, the quicker the water escapes.* This is an astounding but certain fact.

"That deep and distant drains, where a sufficient fall can be obtained, are by far the most profitable, by affording to the roots of the plants a greater range for food."

Of course, where the soil is underlaid by rock, less than four feet from the surface; and where an outlet at that depth cannot be obtained, we must, per force, drain less deeply, but where there exists no such obstacle, drains should be laid at a *general* depth of four feet,—general, not uniform, because the drain should have a uniform inclination, which the surface of the land rarely has.

The Distance between the Drains.—Concerning this, there is less unanimity of opinion among engineers, than prevails with regard to the question of depth.

In tolerably porous soils, it is generally conceded that 40 or even 50 feet is sufficiently near for 4-foot drains, but, for the more retentive clays, all distances from 18 feet to 50 feet are recommended, though those who belong to the more narrow school are, as a rule, extending the limit, as they see, in practice, the complete manner in which drains at wider intervals perform their work. A careful consideration of the experience of the past twenty years, and of the arguments of writers on drainage, leads to the belief that there are few soils, which need draining at all, on which it will be safe to place 4-foot drains at much wider intervals than 40 feet. In the lighter loams there are many instances of the successful application of Professor Mapes' rule, that "3-foot drains should be "placed 20 feet apart, and for each additional foot in "depth the distance may be doubled; for instance, 4-foot "drains should be 40 feet apart, and 5-foot drains 80 feet "apart." But, with reference to the greater distance, (80 feet,) it is not to be recommended in stiff clays, for any depth of drain. Where it is necessary, by reason of insufficient fall, or of underground rock, to go only three feet deep, the drains should be as near together as 20 feet.

At first thought, it may seem akin to quackery to recommend a uniform depth and distance, without reference to the character of the land to be drained; and it is unquestionably true that an exact adaptation of the work to the varying requirements of different soils would be beneficial, though no system can be adopted which will make clay drain as freely as sand. The fact is, that the adjustment of the distances between drains is very far from partaking of the nature of an exact science, and there is really very little known, by any one, of the principles on which it should be based, or of the manner in which the bearing of those principles, in any particular case, is affected by several circumstances which vary with each change of soil, inclination and exposure.

In the essays on drainage which have been thus far published, there is a vagueness in the arguments on this branch of the subject, which betrays a want of definite conviction in the minds of the writers; and which tends quite as much to muddle as to enlighten the ideas of the reader. In so far as the directions are given, whether fortified by argument or not, they are clearly empirical, and are usually very much qualified by considerations which weigh with unequal force in different cases.

In laying out work, any skillful drainer will be guided, in deciding the distance between the lines, by a judgment which has grown out of his former experience; and which will enable him to adapt the work, measurably, to the requirements of the particular soil under consideration; but he would probably find it impossible to so state the reasons for his decision, that they would be of any general value to others.

Probably it will be a long time before rules on this subject, based on well sustained *theory*, can be laid down with distinctness, and, in the mean time, we must be guided by the results of practice, and must confine ourselves to a distance which repeated trial, in various soils, has proven to be safe for all agricultural land. In the drainage of the Central Park, after a mature consideration of all that had been published on the subject, and of a considerable previous observation and experience, it was decided to adopt a general depth of four feet, and to adhere as closely as possible to a uniform distance of forty feet. No instance was known of a failure to produce good results by draining at that distance, and several cases were recalled where drains at fifty and sixty feet had proved so inefficient that intermediate lines became necessary. After from seven to ten years' trial, the Central Park drainage, by its results, has shown that,—although some of the land is of a very retentive character,—this distance is not too great; and it is adopted here for recommendation to all who have no especial reason for supposing that greater distances will be fully effective in their more porous soils.

As has been before stated, drains at that distance, (or at any distance,) will not remove all of the water of saturation from heavy clays so rapidly as from more porous soil; but, although, in some cases, the drainage may be insufficient during the first year, and not absolutely perfect during the second and third years, the increased porosity which drainage causes, (as the summer droughts make fissures in the earth, as decayed roots and other organic deposits make these fissures permanent, and as chemical action in the aërated soil changes its character,) will finally bring clay soils to as perfect a condition as they are capable of attaining, and will invariably render them excellent for cultivation.

The Direction of the Laterals should be *right up and down the slope of the land*, in the line of steepest descent. For a long time after the general adoption of thorough-draining, there was much discussion of this subject, and much variation in practice. The influence of the old rules for making surface or "catch-water" drains lasted for a long time, and there was a general tendency to make tile drains follow the same directions. An important requirement of these was that they should not take so steep an inclination as to have their bottoms cut out and their banks undermined by the rapid flow of water, and that they should arrest and carry away the water flowing down over the surface of hill sides. The arguments for the line of steepest descent were, however, so clear, and drains laid on that line were so universally successful in practice, that it was long ago adopted by all,—save those novices who preferred to gain their education in draining in the expensive school of their own experience.

The more important reasons why this direction is the best are the following: First, it is the quickest way to get the water off. Its natural tendency is to run straight down the hill, and nothing is gained by diverting it from this course. Second, if the drain runs obliquely down the hill, the water will be likely to run out at the joints of the tile and wet the ground below it; even if it do not, mainly, run past the drain from above into the land below, instead of being forced into the tile. Third, a drain lying obliquely across a hillside will not be able to draw the water from below up the hill toward it, and the water of nearly the whole interval will have to seek its outlet through the drain below it. Fourth, drains running directly down the hill will tap any porous water bearing strata, which may crop out, at regular intervals, and will thus prevent the spewing out of the water at the surface, as it might do if only oblique drains ran for a long distance just above or just below them. Very steep, and very springy hill sides, sometimes require very frequent drains to catch the water which has a tendency to flow to the surface; this, however, rarely occurs.

In laying out a plan for draining land of a broken surface, which inclines in different directions, it is impossible to make the drains follow the line of steepest descent, and at the same time to have them all parallel, and at uniform distances. In all such cases a compromise must be made between the two requirements. The more nearly the parallel arrangement can be preserved, the less costly will the work be, while the more nearly we follow the steepest slope of the ground, the more efficient will each drain be. No rule for this adjustment can be given, but a careful study of the plan of the ground, and of its contour lines, will aid in its determination. On all irregular ground it requires great skill to secure the greatest efficiency consistent with economy.

The *fall* required in well made tile drains is very much less than would be supposed, by an inexperienced person, to be necessary. Wherever practicable, without too great cost, it is desirable to have a fall of one foot in one hundred feet, but more than this in ordinary work is not especially to be sought, although there is, of course, no objection to very much greater inclination.

One half of that amount of fall, or six inches in one hundred feet, is quite sufficient, if the execution of the work is carefully attended to.

The least rate of fall which it is prudent to give to a drain, in using ordinary tiles, is 2.5 in 1,000, or three inches in one hundred feet, and even this requires very careful work.* A fall of six inches in one hundred feet is recommended whenever it can be easily obtained—not as being more effective, but as requiring less precision, and consequently less expense.

Kinds and Sizes of Tiles.—Agricultural drain-tiles are made of clay similar to that which is used for brick. When burned, they are from twelve inches to fourteen inches long, with an interior diameter of from one to eight inches, and with a thickness of wall, (depending on the strength of the clay, and the size of the bore,) of from one-quarter of an inch to more than an inch. They are porous, to the extent of absorbing a certain amount of water, but their porosity has nothing to do with their use for drainage,—for this purpose they might as well be of glass. The water enters them, not through their walls,

* Some of the drains in the Central Park have a fall of only 1 in 1,000, and they work perfectly; but they are large mains, laid with an amount of care, and with certain costly precautions, (including precisely graded wooden floors,) which could hardly be expected in private work.

but at their joints, which cannot be made so tight that they will not admit the very small amount of water that will need to enter at each space. Gisborne says:

"If an acre of land be intersected with parallel drains "twelve yards apart, and if on that acre should fall the "very unusual quantity of one inch of rain in twelve "hours, in order that every drop of this rain may be dis- "charged by the drains in forty-eight hours from the com- "mencement of the rain—(and in a less period that quan- "tity neither will, not is it desirable that it should, filter "through an agricultural soil)—the interval between two "pipes will be called upon to pass two-thirds of a table- "spoonful of water per minute, and no more. Inch pipes, "lying at a small inclination, and running only half-full, "will discharge more than double this quantity of water "in forty-eight hours."

Tiles may be made of any desired form of section,—the usual forms are the "horse-shoe," the "sole," the "double-sole," and the "round." The latter may be used with collars, and they constitute the "pipes and collars," frequently referred to in English books on drainage.

Horse-shoe tiles, Fig. 13, are condemned by all modern engineers. Mr. Gisborne disposes of them by an argument of some length, the quotation of which in these pages is probably advisable, because they form so much better conduits than stones, and to that extent have been so successfully employed, that they are still largely used in this country by "amateurs."

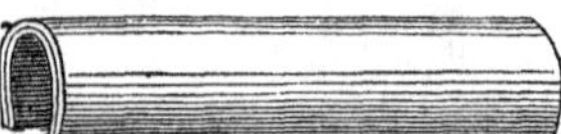
Fig. 13.—HORSE-SHOE TILE.

"We shall shock some and surprise many of our readers, when we state confidently that, in average soils, and, still more, in those which are inclined to be tender, horse shoe tiles form the weakest and most failing conduit which has ever been used for a deep drain. It is so, however; and a little thought, even if we had no experience, will tell us that it must be so. A doggrel song, quite destitute of humor, informs us that tiles of this sort were used in 1760 at Grandesburg Hall, in Suffolk, by Mr. Charles Lawrence, the owner of the estate. The earliest of which we had experience were of large area and of weak form. Constant failures resulted from their use, and the cause was investigated; many of the tiles were found to be choked up with clay, and many to be broken longitudinally through the crown. For the first evil, two remedies were adopted; a sole of slate, of wood, or of its own material, was sometimes placed under the tile, but the more usual practice was to form them with club-feet. To meet the case of longitudinal fracture, the tiles were reduced in size, and very much thickened in proportion to their area. The first of these remedies was founded on an entirely mistaken, and the second on no conception at all of the cause of the evil to which they were respectively applied. The idea was, that this tile, standing on narrow feet, and pressed by the weight of the refilled soil, sank into the floor of the drain; whereas, in fact, the floor of the drain rose into the tile. Any one at all conversant with collieries is aware that when a *strait* work (which is a small subterranean tunnel six feet high and four feet wide or thereabouts) is driven in coal, the rising of the floor is a more usual and far more inconvenient occurrence than the falling of the roof: the weight of the two sides squeezes up the floor. We have seen it formed into a very decided arch without fracture. Exactly a similar operation takes place in the drain. No one had till recently dreamed of forming a tile drain, the bottom of which a man was not to approach personally within twenty inches or two feet. To no one had it then occurred that width at the bottom of the drain was a great evil. For the convenience of the operator the drain was formed with nearly perpendicular sides, of a width in which he could stand and work conveniently, shovel the bottom level with his ordinary spade, and lay the tiles by his hand; the result was a drain with nearly perpendicular sides, and a wide bottom. No sort of clay, particularly when softened by water standing on it or running over it, could fail to rise under such circumstances; and the deeper the drain the greater the pressure and the more certain the rising. A horse-shoe tile, which may be a tolerable secure conduit in a drain of two feet, in one of four feet becomes an almost certain failure. As to the longitudinal fracture—not only is the tile subject to be broken by one of those slips which are so troublesome in deep draining, and to which the lightly-filled material, even when the drain is completed, offers an imperfect resistance, but the constant pressure together of the sides, even when it does not produce a fracture of the soil, catches hold of the feet of the tile, and breaks it through the crown. Consider the case of a drain formed in clay when dry, the conduit a horse-shoe tile. When the clay expands with moisture, it necessarily presses on the tile and breaks it through the crown, its weakest part.* When the Regent's

* The tile has been said, by great authorities, to be broken by contraction, under some idea that the clay envelops the tile and presses it when it contracts. That is nonsense. The contraction would liberate the tile. Drive a stake into wet clay; and when the clay is dry, observe whether it clasps the stake tighter or has released it, and you will no longer have any doubt whether expansion or contraction breaks the tile. Shrink is a better word than contract.

Park was first drained, large conduits were in fashion, and they were made circular by placing one horse-shoe tile upon another. It would be difficult to invent a weaker conduit. On re-drainage, innumerable instances were found in which the upper tile was broken through the crown, and had dropped into the lower. Next came the ⌂ form, tile and sole in one, and much reduced in size—a great advance; and when some skillful operator had laid this tile bottom upwards we were evidently on the eve of pipes. For the ⌂ tile a round pipe moulded with a flat-bottomed solid sole is now generally substituted, and is an improvement; but is not equal to pipes and collars, nor generally cheaper than they are."

One chief objection to the *Sole-tiles* is, that, in the drying which they undergo, preparatory to the burning, the upper side is contracted, by the more rapid drying, and they often require to be trimmed off with a hatchet before they will form even tolerable joints; another is, that they cannot be laid with collars, which form a joint so perfect and so secure, that their use, in the smaller drains, should be considered indispensable.

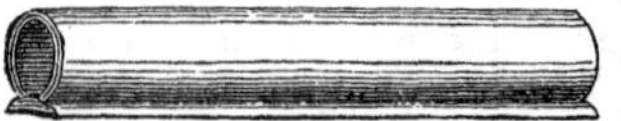
Fig. 14.—SOLE TILE.

The *double-sole tiles*, which can be laid either side up give a much better joint, but they are so heavy as to make the cost of transportation considerably greater. They are also open to the grave objection that they cannot be fitted with collars.

Fig. 15.—DOUBLE-SOLE TILE.

Experience, in both public and private works in this country, and the cumulative testimony of English and French engineers, have demonstrated that the only tile which it is economical to use, is the *best* that can be found, and that the best,—much the best—thus far invented, is the "pipe, or round tile, and collar,"—and these are unhesitatingly recommended for use in all cases. Round tiles of small sizes should not be laid without collars, as the ability to use these constitutes their chief advantage; holding them perfectly in place, preventing the rattling in of loose dirt in laying, and giving twice the space for the entrance of water at the joints. A chief advantage of the larger sizes is, that they may be laid on any side and thus made to fit closely. The usual sizes of these tiles are $1\frac{1}{4}$ inches, $2\frac{1}{4}$ inches, and $3\frac{1}{2}$ inches in interior diameter. Sections of the $2\frac{1}{4}$ inch make collars for the $1\frac{1}{4}$

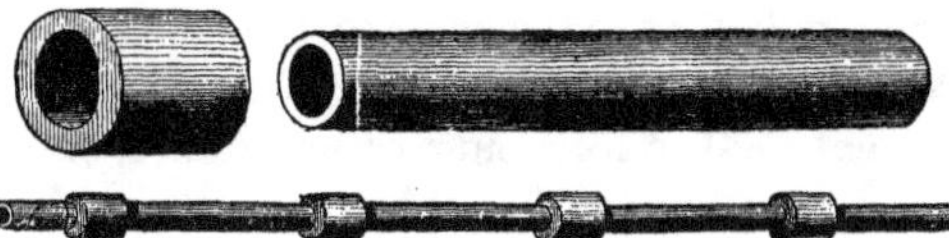

Fig. 16.—ROUND TILE AND COLLAR, AND THE SAME AS LAID.

inch, and sections of the $3\frac{1}{2}$ inch make collars for the $2\frac{1}{4}$ inch. The $3\frac{1}{2}$ inch size does not need collars, as it is easily secured in place, and is only used where the flow of water would be sufficient to wash out the slight quantity of foreign matters that might enter at the joints.

The size of tile to be used is a question of consequence. In England, 1-inch pipes are frequently used, but $1\frac{1}{4}$ inch* are recommended for the smallest drains. Beyond this limit, the proper size to select is, *the smallest that can convey the water which will ordinarily reach it after a heavy rain.* The smaller the pipe, the more concentrated the flow, and, consequently, the more thoroughly obstructions will be removed, and the occasional flushing of the pipe, when it is taxed, for a few hours, to its utmost capacity, will insure a thorough cleansing. No inconvenience can result from the fact that, on rare occasions, the drain is unable, for a short time, to discharge all the water that reaches it, and if collars are used, or if the clay be well packed about the pipes, there need be no fear of the tile being displaced by the pressure. An idea of the drying capacity of a $1\frac{1}{4}$-inch tile may be gained from observing its *wetting* capacity, by connecting a pipe of this size with

* Taking the difference of friction into consideration, $1\frac{1}{4}$ inch pipes have fully twice the discharging capacity of 1-inch pipes.

a sufficient body of water, at its surface, and discharging, over a level dry field, all the water which it will carry. A $1\frac{1}{4}$-inch pipe will remove all the water which would fall on an acre of land in a very heavy rain, in 24 hours,—much less time than the water would occupy in getting to the tile, in any soil which required draining; and tiles of this size are ample for the draining of two acres. In like manner, $2\frac{1}{2}$-inch tile will suffice for eight, and $3\frac{1}{2}$-inch tile for twenty acres. The foregoing estimates are, of course, made on the supposition that only the water which falls on the land, (storm water,) is to be removed. For main drains, when greater capacity is required, two tiles may be laid, (side by side,) or in such cases the larger sizes of sole tiles may be used, being somewhat cheaper. Where the drains are laid 40 feet apart, about 1,000 tiles per acre will be required, and, in estimating the quantity of tiles of the different sizes to be purchased, reference should be had to the following figures; the first 2,000 feet of drains require a collecting drain of $2\frac{1}{4}$-inch tile, which will take the water from 7,000 feet; and for the outlet of from 7,000 to 20,000 feet $3\frac{1}{2}$-inch tile may be used. Collars, being more subject to breakage, should be ordered in somewhat larger quantities.

Of course, such guessing at what is required, which is especially uncertain if the surface of the ground is so irregular as to require much deviation from regular parallel lines, is obviated by the careful preparation of a plan of the work, which enables us to measure, beforehand, the length of drain requiring the different sizes of conduit, and, as tiles are usually made one or two inches more than a foot long, a thousand of them will lay a thousand feet,—leaving a sufficient allowance for breakage, and for such slight deviations of the lines as may be necessary to pass around those stones which are too large to remove. In very stony ground, the length of lines is often materially increased, but in such ground, there is usually rock enough, or such accumulations of boulders in some parts, to reduce the length of drain which it is possible to lay, at least as much as the deviations will increase it.

It is always best to make a contract for tile considerably in advance. The prices which are given in the advertisements of the makers, are those at which a single thousand,—or even a few hundred,—can be purchased, and very considerable reductions of price may be secured on large orders. Especially is this the case if the land is so situated that the tile may be purchased at either one of two tile works,—for the prices of all are extravagantly high, and manufacturers will submit to large discounts rather than lose an important order.

It is especially recommended, in making the contract, to stipulate that every tile shall be hard-burned, and that those which will not give a *clear ring* when struck with a metallic instrument, shall be rejected, and the cost of their transportation borne by the maker. The tiles used in the Central Park drainage were all tested with the aid of a bit of steel which had, at one end, a cutting edge. With this instrument each tile was "sounded," and its hardness was tested by scraping the square edge of the bore. If it did not "ring" when struck, or if the edge was easily cut, it was rejected. From the first cargo there were many thrown out, but as soon as the maker saw that they were really inspected, he sent tile of good quality only. Care should also be taken that no *over-burned* tile,—such as have been melted and warped, or very much contracted in size by too great heat,—be smuggled into the count.

A little practice will enable an ordinary workman to throw out those which are imperfect, and, as a single tile which is so underdone that it will not last, or which, from over-burning, has too small an orifice, may destroy a long drain, or a whole system of drains, the inspection should be thorough.

The collars should be examined with equal care. Concerning the use of these, Gisborne says:

"To one advantage which is derived from the use of "collars we have not yet adverted—the increased facility "with which free water existing in the soil can find en-"trance into the conduit. The collar for a $1\frac{1}{2}$-inch pipe "has a circumference of three inches. The whole space "between the collar and the pipe on each side of the "collar is open, and affords no resistance to the en-"trance of water; while at the same time the superin-"cumbent arch of the collar protects the junction of two "pipes from the intrusion of particles of soil. We con-"fess to some original misgivings that a pipe resting only "on an inch at each end, and lying hollow, might prove "weak and liable to fracture by weight pressing on it "from above; but the fear was illusory. Small particles "of soil trickle down the sides of every drain, and the "first flow of water will deposit them in the vacant space "between the two collars. The bottom, if at all soft, will "also swell up into any vacancy. Practically, if you re-"open a drain well laid with pipes and collars, you will "find them reposing in a beautiful nidus, which, when they "are carefully removed, looks exactly as if it had been "moulded for them."

The cost of collars should not be considered an objection to their use; because, without collars it would not be safe, (as it is difficult to make the orifices of two pieces come exactly opposite to each other,) to use less than 2-inch tiles, while, with collars, $1\frac{1}{4}$-inch are sufficient for the same use, and, including the cost of collars, are hardly more expensive.

It is usual, in all works on agricultural drainage, to insert tables and formulæ for the guidance of those who are to determine the size of tile required to discharge the water of a certain area. The practice is not adopted here, for the reason that all such tables are without practical value. The smoothness and uniformity of the bore; the rate of fall; the depth of the drain, and consequent "head," or pressure, of the water; the different effects of different soils in retarding the flow of the water to the drain; the different degrees to which angles in the line of tile affect the flow; the degree of acceleration of the flow which is caused by greater or less additions to the stream at the junction of branch drains; and other considerations, arising at every step of the calculation, render it impossible to apply delicate mathematical rules to work which is, at best, rude and unmathematical in the extreme. In sewerage, and the water supply of towns, such tables are useful,—though, even in the most perfect of these operations, engineers always make large allowances for circumstances whose influence cannot be exactly measured,—but in land drainage, the ordinary rules of hydraulics have to be considered in so many different bearings, that the computations of the books are not at all reliable. For instance, Messrs. Shedd & Edson, of Boston, have prepared a series of tables, based on Smeaton's experiments, for the different sizes of tile, laid at different inclinations, in which they state that 1½-inch tile, laid with a fall of one foot in a length of one hundred feet, will discharge 12,054.81 gallons of water in 24 hours. This is equal to a rain-fall of over 350 inches per year on an acre of land. As the average annual rain-fall in the United States is about 40 inches, at least one-half of which is removed by evaporation, it would follow, from this table, that a 1½-inch pipe, with the above named fall, would serve for the drainage of about 17 acres. But the calculation is again disturbed by the fact that the rain-fall is not evenly distributed over all the days of the year,—as much as six inches having been known to fall in a single 24 hours, (amounting to about 150,000 gallons per acre,) and the removal of this water in a single day would require a tile nearly five inches in diameter, laid at the given fall, or a 3-inch tile laid at a fall of more than $7\frac{1}{2}$ feet in 100 feet. But, again, so much water could not reach a drain four feet from the surface, in so short a time, and the time required would depend very much on the character of the soil. Obviously, then, these tables are worthless for our purpose. Experience has fully shown that the sizes which are recommended below are ample for practical purposes, and probably the areas to be drained by the given sizes might be greatly increased, especially with reference to such soils as do not allow water to percolate very freely through them.

In connection with this subject, attention is called to the following extract from the Author's Report on the Drainage, which accompanies the "Third Annual Report of the Board of Commissioners of the Central Park:"

"In order to test the efficiency of the system of drainage "employed on the Park, I have caused daily observations "to be taken of the amount of water discharged from the "principal drain of 'the Green,' and have compared it "with the amount of rain-fall. A portion of the record of "those observations is herewith presented.

"In the column headed 'Rain-Fall,' the amount of "water falling on one acre during the entire storm, is given "in gallons. This is computed from the record of a rain-"gauge kept on the Park.

"Under the head of 'Discharge,' the number of gallons "of water drained from one acre during 24 hours is given. "This is computed from observations taken, once a day or "oftener, and supposes the discharge during the entire "day to be the same as at the time of taking the observa-"tions. It is, consequently, but approximately correct:

Date.	Hour.	Rain-fall.	Discharge.	Remarks.
July 13.	10 A. M.	49,916 galls.	184 galls.	Ground dry. No rain since 3d inst.; 2 inches rain fell between 5.15 and 5.45 P. M., and 1-5th of an inch between 5.45 and 7.15.
" 14.	6½ "		4,968 "	
" 15.	6½ "		1,325 "	
" 16.	8 "		1,104 "	
" 16.	6 P. M.	33,398 "	7,764 "	Ground saturated at a depth of 2 feet when this rain commenced.
" 17.			4,319 "	
" 18.	9 A. M.		2,208 "	
" 19.	7 "		1,325 "	
" 20.	6½ "		993 "	
" 21.	11 "		662 "	
" 22.	6½ "		560 "	
" 23.	10 "	1,698 "	515 "	This slight rain only affected the ratio of decrease.
" 24.	7 "		442 "	Nothing worthy of note until Aug. 3.
Aug. 3.	6½ "	8,490 "	191 "	Rain from 3 P. M. to 3.30 P. M.
" 4.	6½ "	13,018 "	184 "	" 4.45 P. M. to 12 M. N.
" 5.	6½ "	45,288 "	368 "	" 12 M. to 6 P. M.
" 5.	6 P. M.		8,280 "	
" 6.	9 A. M.		3,954 "	
" 7.	9 "		2,208 "	
" 8.	6½ "		828 "	
" 9.	6½ "		662 "	
" 12.	6½ "		368 "	Rain 12 M. Aug. 12 to 7 A. M. Aug. 13.
" 13.	7 "	19,244 "	1,104 "	
" 14.	9 "		736 "	
" 24.	9 "	1,132 "	191 "	" 3 A. M. to 4.15 A. M.
" 25.	9 "	5,547 "	9,936 "	" 3.30 P. M. 24th, to 7 A. M. 25th.
" 25.	7 P. M.	566 "	7,740 "	" 7 A. M. to 12 M.
" 26.	6½ A. M.		3,974 "	
" 26.	6 P. M.		2,208 "	
" 27.	6½ A. M.	566 "	1,529 "	" 4 P. M. to 6 P. M.
" 28.	7 "		993 "	
Sep. 11.	7 "	566 "	165 "	" 12 M. N. (10th) to 7 A. M. (11th.)
" 12.	9 "	5,094 "	147 "	" 12 M. (11th) to 7 A. M. (12th.)
" 13.	9 "	566 "	132 "	" 4 P. M. to 6 P. M.
" 16.	9 "	15,848 "	110 "	" 12 M. to 12 M. N.
" 17.	7 "	27,552 "	1,104 "	Rain continued until 12 M.
" 17.	5 P. M.		6,624 "	
" 18.	8 A. M.	566 "	4,968 "	
" 19.	6½ "		2,208 "	
" 19.	4 P. M.		1,805 "	
" 20.	9 A. M.	566 "	1,324 "	Rain f'm 12 M. (19th) to 7 A. M. (20th.)
" 21.	9 "	5,094 "	945 "	" 3.20 P. M. (20th) to 6 A. M. (21st.)
" 22.	9 "	10,185 "	1,656 "	" 12 M. (21st) to 7 A. M. (22d.)
" 23.	9 "	40,756 "	7,948 "	Rain continued until 7 A. M. (23d.)
" 24.	9 "		4,968 "	
" 25.	9 "	566 "	2,984 "	
" 26.	9 "		2,484 "	
Oct. 1.	9 "		828 "	There was not enough rain during this period to materially affect the flow of water.
Nov. 18.	9 "		83 "	
" 19.	9 "	1,132 "	184 "	Rain 4.50 P. M. (18th) to 8 A. M. (19th.)
" 20.	9 "		119 "	
" 22.	9 "	29,336 "	6,624 "	Rain all of the previous night.
" 22.	2 P. M.		6,624 "	
" 23.	9 A. M.		4,968 "	
" 24.	9 "		1,711 "	
" 24.	2 P. M.		1,417 "	
Dec. 17.	9 A. M.		552 "	
" 18.	9 "		4,968 "	Rain during the previous night.
" 30.	10 "		581 "	

"The tract drained by this system, though very swampy, "before being drained, is now dry enough to walk upon, "almost immediately after a storm, except when underlaid "by a stratum of frozen ground."

The area drained by the main at which these gaugings were made, is about ten acres, and, in deference to the prevailing mania for large conduits, it had been laid with 6-inch sole-tile. The greatest recorded discharge in 24 hours was (August 25th,) less than 100,000 gallons from the ten acres,—an amount of water which did not half fill the tile, but which, according to the tables referred to, would have entirely filled it.

In view of all the information that can be gathered on the subject, the following directions are given as perfectly reliable for drains four feet or more in depth, laid on a well regulated fall of even three inches in a hundred feet:

For	2 acres	1¼	inch	pipes	(with collars.)
For	8 acres	2¼	"	"	(" ")
For	20 acres	3½	"	"	
For	40 acres	2 3½	"	"	or one 5-inch sole-tile.
For	50 acres	6	"	"	sole-tile.
For	100 acres	8	"	"	or two 6-inch sole-tiles.

It is not pretended that these drains will immediately remove all the water of the heaviest storms, but they will always remove it fast enough for all practical purposes, and, if the pipes are securely laid, the drains will only be benefited by the occasional cleansing they will receive when running "more than full." In illustration of this statement, the following is quoted from a paper communicated by Mr. Parkes to the Royal Agricultural Society of England in 1843:

"Mr. Thomas Hammond, of Penshurst, (Kent,) now "uses no other size for the parallel drains than the inch "tile in the table, (No. 5,) having commenced with No.

" 4,* and it may be here stated, that the opinion of all the " farmers who have used them in the Weald, is that a bore " of an inch area is abundantly large. A piece of 9 acres, " now sown with wheat, was observed by the writer, 36 " hours after the termination of a rain which fell heavily " and incessantly during 12 hours on the 7th of Novem- " ber. This field was drained in March, 1842, to the depth " of 30 to 36 inches, at a distance of 24 feet asunder, the " length of each drain being 235 yards.

"Each drain emptied itself through a fence bank into " a running stream in a road below it; the discharge " therefore was distinctly observable. Two or three of " the pipes had now ceased running; and, with the ex- " ception of one which tapped a small spring and gave a " stream about the size of a tobacco pipe, the run from " the others did not exceed the size of a wheat straw. " The greatest flow had been observed by Mr. Hammond " at no time to exceed half the bore of the pipes. The " fall in this field is very great, and the drains are laid in " the direction of the fall, which has always been the prac- " tice in this district. The issuing water was transpa- " rently clear; and Mr. Hammond states that he has " never observed cloudiness, except for a short time after " very heavy flushes of rain, when the drains are quickly " cleared of all sediment, in consequence of the velocity " and force of the water passing through so small a channel. " Infiltration through the soil and into the pipes, must, " in this case, be considered to have been perfect; and " their observed action is the more determinate and valua- " ble as regards time and effect, as the land was saturated " with moisture previous to this particular fall of rain, " and the pipes had ceased to run when it commenced. " This piece had, previous to its drainage, necessarily " been cultivated in narrow stretches, with an open water

* No. 5 was one inch in diameter; No. 4, about 1⅛ inches.

" furrow between them; but it was now laid quite plain, " by which one-eighth of the continuation of acreage has " been saved. Not, however, being confident as to the " soil having already become so porous as to dispense en- " tirely with surface drains, Mr. Hammond had drawn " two long water furrows diagonally across the field. On " examining these, it appeared that very little water had " flowed along any part of them during these 12 hours of " rain,—no water had escaped at their outfall; the entire " body of rain had permeated the mass of the bed, and " passed off through the inch pipes; no water perceptible " on the surface, which used to carry it throughout. The " subsoil is a brick clay, but it appears to crack very " rapidly by shrinkage consequent to drainage."

Obstructions.—The danger that drains will become obstructed, if not properly laid out and properly made, is very great, and the cost of removing the obstructions, (often requiring whole lines to be taken up, washed, and relaid with the extra care that is required in working in old and soft lines,) is often greater than the original cost of the improvement. Consequently, the possibility of tile drains becoming stopped up should be fully considered at the outset, and every precaution should be taken to prevent so disastrous a result.

The principal causes of obstruction are *silt*, *vermin*, and *roots*.

Silt is earth which is washed into the tile with the water of the soil, and which, though it may be carried along in suspension in the water, when the fall is good, will be deposited in the eddies and slack-water, which occur whenever there is a break in the fall, or a defect in the laying of the tile.

Whenever it is possible to avoid it, no drain should have a decreasing rate of fall as it approaches its outlet.

If the first hundred feet from the upper end of the

drain has a fall of three inches, the next hundred feet should not have less than three inches, lest the diminished velocity cause silt, which required the speed which that fall gives for its removal, to be deposited and to choke the tile. This defect of grade is shown in Fig. 17. If the second hundred feet has an inclination of *more* than three inches, (Fig. 18,) the removal of silt will be even better secured than if the fall continued at the original rate. Some silt will enter newly made drains, in spite of our utmost care, but the amount should be very slight, and if it is evenly deposited throughout the whole length of the drain, (as it sometimes is when the rate of fall is very low,) it will do no especial harm; but it becomes dangerous when it is accumulated within a short distance, by a decreasing fall, or by a single badly laid tile, or imperfect joint, which, by arresting the flow, may cause as much mischief as a defective grade.

Owing to the general conformation of the ground, it is sometimes absolutely necessary to adopt such a grade as is shown in Fig. 19,—even to the extent of bringing the drain down a rapid slope, and continuing it with the least possible fall through level ground. When such changes must be made, they should be effected by angles, and not by curves. In *increasing* the fall, curves in the grade are always advisable, in *decreasing* it they are always objectionable, except when the decreased fall is still considerable,—say, at least 2 feet in 100 feet. The reason for making an absolute angle at the point of depression is, that it enables us to catch the silt at that point in a silt basin, from which it may be removed as occasion requires.

A Silt Basin is a chamber, below the grade of the drain, into which the water flows, becomes comparatively quiet, and deposits its silt, instead of carrying it into the tile beyond. It may be large or small, in proportion to the amount of drain above, which it has to accommodate. For a few hundred feet of the smallest tile, it may be only a

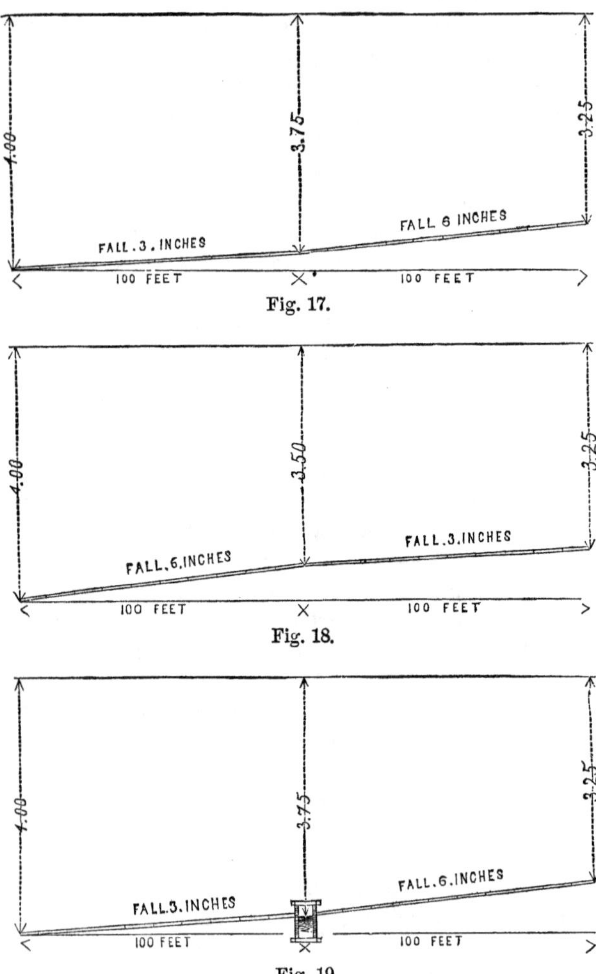

Fig. 17.

Fig. 18.

Fig. 19.

THREE PROFILES OF DRAINS, WITH DIFFERENT INCLINATIONS.

6-inch tile placed on end and sunk so as to receive and discharge the water at its top. For a large main, it may be a brick reservoir with a capacity of 2 or 3 cubic feet. The position of a silt basin is shown in Fig. 19.

The quantity of silt which enters the drain depends very much on the soil. Compact clays yield very little, and wet, running sands, (quicksands,) a great deal. In a soil of the latter sort, or one having a layer of running sand at the level of the drain, the ditch should be excavated a little below the grade of the drain, and then filled to that level with a retentive clay, and rammed hard. In all cases when the tile is well laid, (especially if collars are used,) and a stiff earth is well packed around the tile, silt will not enter the drain to an injurious extent, after a few months' operation shall have removed the loose particles about the joints, and especially after a few very heavy rains, which, if the tiles are small, will sometimes wash them perfectly clean, although they may have been half filled with dirt.

Vermin,—field mice, moles, etc.,—sometimes make their nests in the tile and thus choke them, or, dying in them, stop them up with their carcases. Their entrance should be prevented by placing a coarse wire cloth or grating in front of the outlets, which afford the only openings for their entrance.

Roots.—The roots of many water-loving trees,—especially willows,—will often force their entrance into the joints of the tile and fill the whole bore with masses of fibre which entirely prevent the flow of water. Collars make it more difficult for them to enter, but even these are not a sure preventive. Gisborne says:

"My own experience as to roots, in connection with "deep pipe draining, is as follows: I have never known "roots to obstruct a pipe through which there was not a "perennial stream. The flow of water in summer and "early autumn appears to furnish the attraction. I have

" never discovered that the roots of any esculent vegetable " have obstructed a pipe. The trees which, by my own " personal observation, I have found to be most danger- " ous, have been red willow, black Italian poplar, alder, " ash, and broad-leaved elm. I have many alders in close " contiguity with important drains, and, though I have " never convicted one, I cannot doubt that they are dan- " gerous. Oak, and black and white thorns, I have not " detected, nor do I suspect them. The guilty trees have " in every instance been young and free growing; I have " never convicted an adult. These remarks apply solely " to my own observation, and may of course be much " extended by that of other agriculturists. I know an in- " stance in which a perennial spring of very pure and (I " believe) soft water is conveyed in socket pipes to a " paper mill. Every junction of two pipes is carefully " fortified with cement. The only object of cover being " protection from superficial injury and from frost, the " pipes are laid not far below the sod. Year by year these " pipes are stopped by roots. Trees are very capricious in " this matter. I was told by the late Sir R. Peel that he " sacrificed two young elm trees in the park at Drayton " Manor to a drain which had been repeatedly stopped by " roots. The stoppage was nevertheless repeated, and " was then traced to an elm tree far more distant than " those which had been sacrificed. Early in the autumn " of 1850 I completed the drainage of the upper part of a " boggy valley, lying, with ramifications, at the foot of " marly banks. The main drains converge to a common " outlet, to which are brought one 3-inch pipe and three of 4 " inches each. They lie side by side, and water flows pe- " rennially through each of them. Near to this outlet did " grow a red willow. In February, 1852, I found the " water breaking out to the surface of the ground about " 10 yards above the outlet, and was at no loss for the " cause, as the roots of the red willow showed themselves

" at the orifice of the 3-inch and of two of the 4-inch pipes. " On examination I found that a root had entered a joint " between two 3-inch pipes, and had traveled 5 yards to " the mouth of the drain, and 9 yards up the stream, " forming a continuous length of 14 yards. The root which " first entered had attained about the size of a lady's little " finger; and its ramifications consisted of very fine and " almost silky fibres, and would have cut up into half a " dozen comfortable boas. The drain was completely " stopped. The pipes were not in any degree displaced. " Roots from the same willow had passed over the 3-inch " pipes, and had entered and entirely stopped the first " 4-inch drain, and had partially stopped the second. At " a distance of about 50 yards a black Italian poplar, " which stood on a bank over a 4-inch drain, had com- " pletely stopped it with a bunch of roots. The whole of " this had been the work of less than 18 months, including " the depth of two winters. A 3-inch branch of the same " system runs through a little group of black poplars. " This drain conveys a full stream in plashes of wet, and " some water generally through the winter months, but " has not a perennial flow. I have perceived no indica- " tion that roots have interfered with this drain. I draw " no general conclusions from these few facts, but they " may assist those who have more extensive experience in " drawing some, which may be of use to drainers."

Having considered some of the principles on which our work should be based, let us now return to the map of the field, and apply those principles in planning the work to be done to make it dry.

The Outlet should evidently be placed at the present point of exit of the brook which runs from the springs, collects the water of the open ditches, and spreads over the flat in the southwest corner of the tract, converting it into a swamp. Suppose that, by going some distance into the next field, we can secure an outlet of 3 feet and

9 inches (3.75) below the level of the swamp, and that we decide to allow 3 inches drop between the bottom of the tile at that point, and the reduced level of the brook to secure the drain against the accumulation of sand, which might result from back water in time of heavy rain. This fixes the depth of drain at the outlet at $3\frac{1}{2}$ (3.50) feet.

At that side of the swamp which lies nearest to the main depression of the up-land, (See Fig. 21,) is the proper place at which to collect the water from so much of the field as is now drained by the main brook, and at that point it will be well to place a *silt basin* or well, built up to the surface, which may, at any time, be uncovered for an observation of the working of the drains. The land between this point and the outlet is absolutely level, requiring the necessary fall in the drain which connects the two, to be gained by raising the upper end of it. As the distance is nearly 200 feet, and as it is advisable to give a fall at least five-tenths of a foot per hundred feet to so important an outlet as this, the drain at the silt basin may be fixed at only $2\frac{1}{2}$ feet. The basin being at the foot of a considerable rise in the ground, it will be easy, within a short distance above, to carry the drains which come to it to a depth of 4 feet,—were this not the case, the fall between the basin and the outlet would have to be very much reduced.

Main Drains.—The valley through which the brook now runs is about 80 feet wide, with a decided rise in the land at each side. If one main drain were laid in the center of it, all of the laterals coming to the main would first run down a steep hillside, and then across a stretch of more level land, requiring the grade of each lateral to be broken at the foot of the hill, and provided with a silt basin to collect matters which might be deposited when the fall becomes less rapid. Consequently, it is best to provide two mains, or collecting drains, (*A* and *C*,) one lying at the foot of each hill, when they will receive the laterals at their greatest fall; but, as these are too far apart to completely drain the valley between them, and are located on land higher than the center of the valley, a drain, (*B*,) should be run up, midway between them.

The collecting drain, *A*, will receive the laterals from the hill to the west of it, as far up as the 10-foot contour line, and, above that point,—running up a branch of the valley, —it will receive laterals from both sides. The drain, *B*, may be continued above the dividing point of the valley, and will act as one of the series of laterals. The drain, *C*, will receive the laterals and sub-mains from the rising ground to the east of it, and from both sides of the minor valley which extends in that direction.

Most of the valley which runs up from the easterly side of the swamp must be drained independently by the drain *E*, which might be carried to the silt basin, did not its continuation directly to the outlet offer a shorter course for the removal of its water. This drain will receive laterals from the hill bordering the southeasterly side of the swamp, and, higher up, from both sides of the valley in which it runs.

In laying out these main drains, more attention should be given to placing them where they will best receive the water of the laterals, and on lines which offer a good and tolerably uniform descent, than to their use for the immediate drainage of the land through which they pass. Afterward, in laying out the laterals, the use of these lines as local drains should, of course, be duly considered.

The Lateral Drains should next receive attention, and in their location and arrangement the following rules should be observed:

1st. They should run down the steepest descent of the land.

2d. They should be placed at intervals proportionate to their depth;—if 4 feet deep, at 40 feet intervals; if 3 feet deep, at 20 feet intervals.

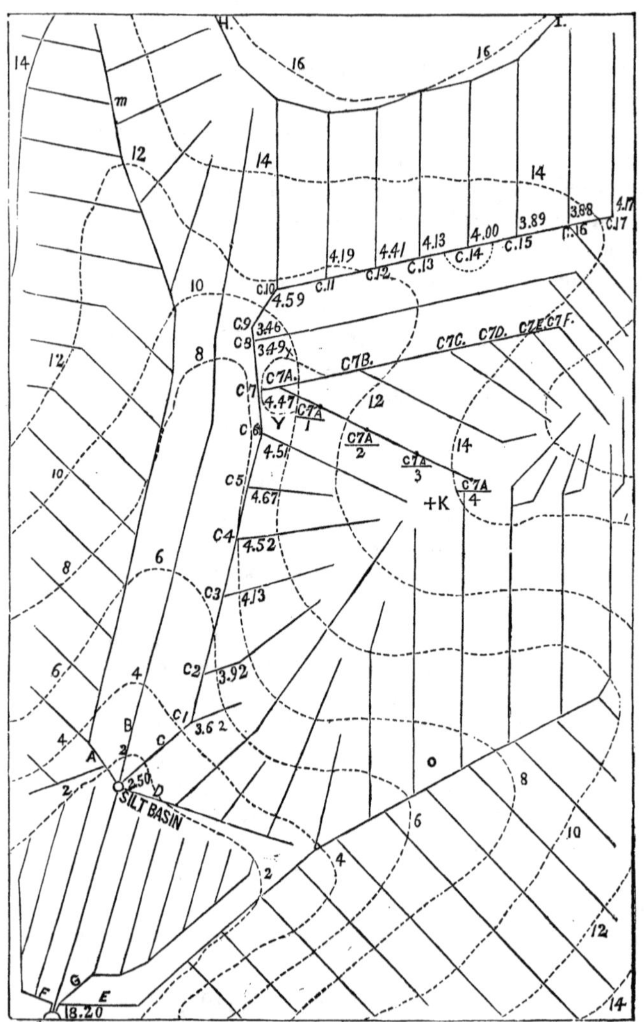

Fig. 20.—MAP WITH DRAINS AND CONTOUR LINES.

3d. They should, as nearly as possible, run parallel to each other.

On land of perfectly uniform character, (all sloping in the same direction,) all of these requirements may be complied with, but on irregular land it becomes constantly necessary to make a compromise between them. Drains running down the line of steepest descent cannot be parallel,—and, consequently, the intervals between them cannot be always the same; those which are farther apart at one end than at the other cannot be always of a depth exactly proportionate to their intervals.

In the adjustment of the lines, so as to conform as nearly to these requirements as the shape of the ground will allow, there is room for the exercise of much skill, and on such adjustment depend, in a great degree, the success and economy of the work. Remembering that on the map, the line of steepest descent is exactly perpendicular to the contour lines of the land, it will be profitable to study carefully the system of drains first laid out, erasing and making alterations wherever it is found possible to simplify the arrangement.

Strictly speaking, all *angles* are, to a certain extent, wasteful, because, if two parallel drains will suffice to drain the land between them, no better drainage will be effected by a third drain running across that land. Furthermore, the angles are practically supplied with drains at less intervals than are required,—for instance, at *C* 7 *a* on the map the triangles included within the dotted line *x*, *y*, will be doubly drained. So, also, if any point of a 4-foot drain will drain the land within 20 feet of it, the land included within the dotted line forming a semi-circle about the point *C* 14, might drain into the end of the lateral, and it no more needs the action of the main drain than does that which lies between the laterals. Of course, angles and connecting lines are indispensable, except where the laterals can run independently across the entire field, and discharge beyond it. The longer the laterals can be made, and the more angles can be avoided, the more economical will the arrangement be ; and, until the arrangement of the lines has been made as nearly perfect as possible, the time of the drainer can be in no way so profitably spent as in amending his plan.

The series of laterals which discharge through the mains *A*, *C*, *D* and *E*, on the accompanying map, have been very carefully considered, and are submitted to the consideration of the reader, in illustration of what has been said above.

At one point, just above the middle of the east side of the field, the laterals are placed at a general distance of 20 feet, because, as will be seen by reference to Fig. 4, a ledge of rock, underground, will prevent their being made more than 3 feet deep.

The line from *H* to *I*, (Fig. 20,) at the north side of the field, connecting the heads of the laterals, is to be a stone and tile drain, such as is described on page 60, intended to collect the water which follows the surface of the rock. (See Fig. 4.)

The swamp is to be drained by itself, by means of two series of laterals discharging into the main lines *F* and *G*, which discharge at the outlet, by the side of the main drain from the silt-basin. By this arrangement, these laterals, especially at the north side of the swamp, being accurately laid, with very slight inclinations, can be placed more deeply than if they ran in an east and west direction, and discharged into the main, which has a greater inclination, and is only two and a half feet deep at the basin. Being $3\frac{1}{2}$ (3.50) feet deep at the outlet, they may be made fully 3 feet deep at their upper ends, and, being only 20 feet apart, they will drain the land as well as is possible. The drains being now laid out, over the whole field, the next thing to be attended to is

The Ordering of the Tile.—The main line from the outlet up to the silt-basin, should be of 3½-inch tiles, of which about 190 feet will be required. The main drain *A* should be laid with 2¼-inch tiles to the point marked *m*, near its upper end, as the lateral entering there carries the water of a spring, which is supposed to fill a 1¼-inch tile. The length of this drain, from the silt-basin to that point is 575 feet. The main drain *C* will require 2¼ inch tiles from the silt-basin to the junction with the lateral, which is marked *C* 10, above which point there is about 1,700 feet of drain discharging into it, a portion of which, being a stone-and-tile drain at the foot of a rock, may be supposed to receive more water than that which lies under the rest of the land;—distance 450 feet. The main drain *E* requires 2¼-inch tiles from the outlet to the point marked *o*, a distance of 380 feet. This tile will, in addition to its other work, carry as much water from the spring, on the line of its fourth lateral, as would fill a 1¼-inch pipe.*

The length of the main drains above the points indicated, and of all the laterals, amounts to about 12,250 feet. These all require 1¼-inch tiles.

Allowing about five per cent. for breakage, the order in round numbers, will be as follows: †

3½-inch round tiles	- - - - - -	200	feet.
2¼ " " "	- - - - - - -	1,500	"
1¼ " " "	- - - - - - -	13,000	"
3½ " Collars	- - - - - -	1,600	
2¼ " "	- - - - - - -	13,250	

* If the springs, when running at their greatest volume, be found to require more than 1¼-inch tiles, due allowance must be made for the increase.

† Owing to the irregularity of the ground, and the necessity for placing some of the drains at narrower intervals, the total length of tile exceeds by nearly 50 per cent. what would be required if it had a uniform slope, and required no collecting drains. It is much greater than will be required in any ordinary case, as a very irregular surface has been adopted here for purposes of illustration.

Order, also, 25 6-inch sole-tiles, to be used in making small silt-basins.

It should be arranged to have the tiles all on the ground before the work of ditching commences, so that there may be no delay and consequent danger to the stability of the banks of the ditches, while waiting for them to arrive. As has been before stated, it should be especially agreed with the tile-maker, at the time of making the contract, that every tile should be perfect;—of uniform shape, and neither too much nor too little burned.

Staking Out.—Due consideration having been given to such preliminaries as are connected with the mapping of the ground, and the arrangement, on paper, of the drains to be made, the drainer may now return to his field, and, while awaiting the arrival of his tiles, make the necessary preparation for the work to be done. The first step is to fix certain prominent points, which will serve to connect the map with the field, by actual measurements, and this will very easily be done by the aid of the stakes which are still standing at the intersections of the 50-foot lines, which were used in the preliminary levelling.

Commencing at the southwest corner of the field, and measuring toward the east a distance of 34 feet, set a pole to indicate the position of the outlet. Next, mark the center of the silt-basin at the proper point, which will be found by measuring 184 feet up the western boundary, and thence toward the east 96 feet, on a line parallel with the nearest row of 50-foot stakes. Then, in like manner, fix the points *C* 1, *C* 6, *C* 9, *C* 10, and *C* 17, and the angles of the other main lines, marking the stakes, when placed, to correspond with the same points on the map. Then stake the angles and the upper ends of the laterals, and mark these stakes to correspond with the map.

It will greatly facilitate this operation, if the plan of the drains which is used in the field, from which the horizontal lines should be omitted, have the intersecting 50-foot lines drawn upon it, so that the measurements may be made from the nearest points of intersection.*

Having staked these guiding points of the drains, it is advisable to remove all of the 50-foot stakes, as these are of no further use, and would only cause confusion. It will now be easy to set the remaining stakes,—placing one at every 50 feet of the laterals, and at the intersections of all the lines.

A system for marking the stakes is indicated on the map, (in the *C* series of drains,) which, to avoid the confusion which would result from too much detail on such a small scale, has been carried only to the extent necessary for illustration. The stakes of the line *C* are marked *C* 1, *C* 2, *C* 3, etc. The stakes of the sub-main *C* 7, are marked *C* 7*a*, *C* 7*b*, *C* 7*c*, etc. The stakes of the lateral which enters this drain at *C* 7*a*, are marked $\frac{C\,7a}{1}$, $\frac{C\,7a}{2}$, $\frac{C\,7a}{3}$, etc. etc. This system, which connects the lettering of each lateral with its own sub-main and main, is perfectly simple, and avoids the possibility of confusion. The position of the stakes should all be lettered on the map, at the original drawing, and the same designating marks put on the stakes in the field, as soon as set.

Grade Stakes, (pegs about 8 or 10 inches long,) should be placed close at the sides of the marked stakes, and driven nearly their full length into the ground. The tops of these stakes furnish fixed points of elevation from which to take the measurements, and to make the computations necessary to fix the depth of the drain at each stake. If the measurements were taken from the surface of the ground, a slight change of position in placing the instrument, would often make a difference of some inches in the depth of the drain.

* The stakes used may be 18 inches long, and driven one-half of their length into the ground. They should have one side sufficiently smooth to be distinctly marked with red chalk.

Taking the Levels.—For accurate work, it is necessary to ascertain the comparative levels of the tops of all of the grade stakes; or the distance of each one of them below an imaginary, horizontal plane. This plane, (in which we use only such lines as are directly above the drains,) may be called the "Datum Line." Its elevation should be such that it will be above the highest part of the land, and, for convenience, it is fixed at the elevation of the levelling instrument when it is so placed as to look over the highest part of the field.

Levelling Instruments are of various kinds. The best for the work in hand, is the common railroad level, which is shown in Fig. 6. This is supported on three legs, which bring it to about the level of the eye. Its essential parts are a telescope, which has two cross-hairs intersecting each other in the line of sight, and which may be turned on its pivot toward any point of the horizon; a bubble glass placed exactly parallel to the line of sight, and firmly secured in its position so as to turn with the telescope; and an apparatus for raising or depressing any side of the instrument by means of set-screws. The instrument is firmly screwed to the tripod, and placed at a point convenient for looking over a considerable part of the highest land. By the use of the set-screws, the plane in which the instrument revolves is brought to a level, so that in whatever direction the instrument is pointed, the bubble will be in the center of the glass. The line of sight, whichever way it is turned, is now in our imaginary plane. A convenient position for the instrument in the field under consideration, would be at the point, east of the center, marked *K*, which is about 3 feet below the level of the highest part of the ground. The telescope should stand about 5 feet above the surface of the ground directly under it.

The Levelling-Rod, (See Fig. 7,) is usually 12 feet long, is divided into feet and hundredths of a foot, and has a movable target which may be placed at any part of its entire length. This is carried by an attendant, who holds it perpendicularly on the top of the grade-stake, while the operator, looking through the telescope, directs him to move the target up and down until its center is exactly in the line of sight. The attendant then reads the elevation, and the operator records it as the distance below the *datum-line* of the top of the grade-stake. For convenience, the letterings of the stakes should be systematically entered in a small field book, before the work commences, and this should be accompanied by such a sketch of the plan as will serve as a guide to the location of the lines on the ground.

The following is the form of the field book for the main drain *C*, with the levels recorded:

Lettering of the Stake.	Depth from Datum Line.
Silt Basin	18.20
C 1	15.44
C 2	14.36
C 3	12.85
C 4	12.18
C 5	11.79
C 6	11.69
C 7	11.55
C 8	11.37
C 9	11.06
C 10	8.94
C 11	8.52
C 12	7.86
C 13	7.70
C 14	7.39
C 15	7.06
C 16	6.73
C 17	5.90

The levelling should be continued in this manner, until the grades of all the points are recorded in the field book.

If, from too great depression of the lower parts of the field, or too great distances for observation, it becomes necessary to take up a new position with the instrument, the new level should be connected, by measurement, with

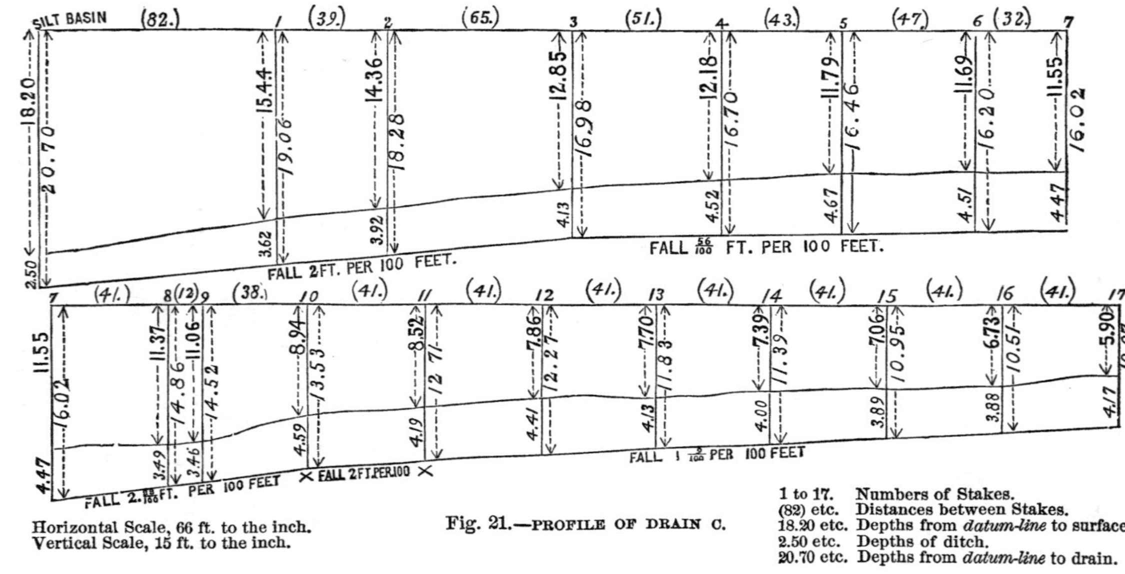

Horizontal Scale, 66 ft. to the inch.
Vertical Scale, 15 ft. to the inch.

Fig. 21.—PROFILE OF DRAIN C.

1 to 17. Numbers of Stakes.
(82) etc. Distances between Stakes.
18.20 etc. Depths from *datum-line* to surface.
2.50 etc. Depths of ditch.
20.70 etc. Depths from *datum-line* to drain.

the old one, and the new observations should be computed to the original plane.

It is not necessary that these levels should be noted on the map,—they are needed only for computing the depth of cutting, and if entered on the map, might be mistaken for the figures indicating the depth, which it is more important to have recorded in their proper positions, for convenience of reference during the work.

The Depth and Grade of the Drains.—Having now staked out the lines upon the land, and ascertained and recorded the elevations at the different stakes, it becomes necessary to determine at what depth the tile shall be placed at each point, so as to give the proper fall to each line, and to bring all of the lines of the system into accord. As the simplest means of illustrating the principle on which this work should be done, it will be convenient to go through with the process with reference to the main drain *C*, of the plan under consideration. A profile of this line is shown in Fig. 21, where the line is broken at stake No. 7, and continued in the lower section of the diagram. The topmost line, from "Silt Basin" to "17," is the horizontal datum-line. The numbers above the vertical lines indicate the stakes; the figures in brackets between these, the number of feet between the stakes; and the heavy figures at the left of the vertical lines, the recorded measurements of depth from the datum-line to the surface of the ground, which is indicated by the irregular line next below the datum-line. The vertical measurements are, of course, very much exaggerated, to make the profile more marked, but they are in the proper relation to each other.

The depth at the silt-basin is fixed at 2½ feet (2.50.) The rise is rapid to stake 3, very slight from there to stake 7, very rapid from there to stake 10, a little less rapid from there to stake 11, and still less rapid from there to stake 17.

To establish the grade by the profile alone, the proper

course would be to fix the depth at the stakes at which the inclination is to be changed, to draw straight lines between the points thus found, and then to measure the vertical distance from these lines to the line indicating the surface of the ground at the different stakes; thus, fixing the depth at stake 3, at 4 feet and 13 hundredths,* the line drawn from that point to the depth of 2.50, at the silt-basin, will be 3 feet and 62 hundredths (3.62) below stake 1, and 3 feet and 92 hundredths (3.92) below stake 2. At stake 7 it is necessary to go sufficiently deep to pass from 7 to 10, without coming too near the surface at 9, which is at the foot of a steep ascent. A line drawn straight from 4.59 feet below stake 10 to 4.17 feet at stake 17, would be unnecessarily deep at 11, 12, 13, and 14; and, consequently it is better to rise to 4.19 feet at 11. So far as this part of the drain is concerned, it would be well to continue the same rise to 12, but, in doing so, we would come too near the surface at 13, 14, and 15; or must considerably depress the line at 16, which would either make a bad break in the fall at that point, or carry the drain too deep at 17.

By the arrangement adopted, the grade is broken at 3, 7, 10, and 11. Between these points, it is a straight line, with the rate of fall indicated in the following table, which commences at the upper end of the drain and proceeds toward its outlet:

From Stake,	Depth,	To Stake,	Depth.	Distance.	Total Fall.	Rate of Fall Per 100 Feet.
No. 17..	4.17 feet.	No. 11..	4.19 feet.	246 feet.	2.64 feet.	1.09 feet.
No. 11..	4.19 "	No. 10..	4.59 "	41 "	82 "	2.00 "
No. 10..	4.59 "	No. 7..	4.47 "	91 "	2.49 "	2.83 "
No. 7..	4.47 "	No. 3..	4.13 "	173 "	96 "	56 "
No. 3..	4.13 "	S. Basin	2.25 "	186 "	3.47 "	1.87 "

It will be seen that the fall becomes more rapid as we ascend from stake 7, but below this point it is very much

* The depth of 4.13, in Fig. 21, as well as the other depths at the points at which the grade changes, happen to be those found by the computation, as hereafter described, and they are used here for illustration.

reduced, so much as to make it very likely that silt will be deposited, (see page 91), and the drain, thereby, obstructed. To provide against this, a silt-basin must be placed at this point which will collect the silt and prevent its entrance into the more nearly level tile below. The construction of this silt-basin is more particularly described in the next chapter. From stake 7 to the main silt-basin the fall is such that the drain will clear itself.

The drawing of regular profiles, for the more imporant drains, will be useful for the purpose of making the beginner familiar with the method of grading, and with the principles on which the grade and depth are computed; and sometimes, in passing over very irregular surfaces, this method will enable even a skilled drainer to hit upon the best adjustment in less time than by computation. Ordinarily, however, the form of computation given in the following table, which refers to the same drain, (*C*,) will be more expeditious, and its results are mathematically more correct.*

No. of Stake.	*Distance Between Stakes.*	*Fall. Feet and Decimals.*		*Depth from Datum Line.*		*Depth of Drain.*	*Remarks.*
		Per 100 Feet.	*Bet'wn Stakes.*	*To Drain.*	*To Surface.*		
Silt Basin.				20.70 ft.	18.20 ft.	2 50 ft.	
C. 1.	82 ft.	2 ft.	1.64 ft.	19.06 "	15.44 "	3.48 "	
C. 2.	39 "	do.	.78 "	18.28 "	14.36 "	3.83 "	
C. 3.	65 "	do.	1.30 "	16.98 "	12.85 "	4 13 "	
C. 4.	51 "	.56	.28 "	16.70 "	12.18 "	4.52 "	
C. 5.	43 "	do.	.24 "	16.46 "	11.79 "	4.67 "	
C. 6.	47 "	do.	.26 "	16.20 "	11.69 "	4.51 "	
C. 7.	32 "	do.	.18 "	16.02 "	11.55 "	4 47 "	Silt-Basin here.
C. 8.	41 "	2.83	1.16 "	14.86 "	11.37 "	3.49 "	Made deep at Nos. 7 and 10 to pass a depression of the surface at No. 9.
C. 9.	12 "	do.	.34 "	14.52 "	11.06 "	3.46 "	
C. 10.	38 "	do.	.99 "	13 53 "	8.94 "	4.59 "	
C. 11.	41 "	2.00	.82 "	12.61 "	8.52 "	4.19 "	
C. 12.	41 "	1.09	.44 "	12.27 "	7.86 "	4.41 "	
C. 13.	41 "	do.	.44 "	11.83 "	7.70 "	4.13 "	
C. 14.	41 "	do.	,44 "	11.39 "	7.39 "	4.00 "	
C. 15.	41 "	do.	.44 "	10.95 "	7.06 "	3.89 "	
C. 16.	41 "	do.	.41 "	10 51 "	6.73 "	3.88 "	
C. 17.	41 "	do.	.44 "	10.07 "	5.90 "	4.17 "	

* The figures in this table, as well as in the next preceding one, are adopted for the published profile of drain *C*, Fig. 21, to avoid confusion. In ordinary cases, the points which are fixed as the basis of the computation are given in round numbers;—for instance, the depth at *C*3 would be assumed to be 4.10 or 4.20, instead of 4.13. The fractions given in the table, and in Fig. 21, arise from the fact that the decimals are not absolutely correct, being carried out only for two figures.

NOTE.—The method of making the foregoing computation is this:

1st. Enter the lettering of the stakes in the first column, commencing at the lower end of the drain.

2d. Enter the distances between each two stakes in the second column, placing the measurement on the line with the number of the *upper* stake of the two.

3d. In the next to the last column enter, on the line with each stake, its depth below the datum-line, as recorded in the field book of levels, (See page 105.)

4th. On the first line of the last column, place the depth of the lower end of the drain, (this is established by the grade of the main or other outlet at which it discharges.)

5th. Add this depth to the first number of the line next preceding it, and enter the sum obtained on the first line of the fifth column, as the depth of the *drain* below the datum-line.

6th. Having reference to the grade of the surface, (as shown by the figures in the sixth column,) as well as to any necessity for placing the drain at certain depths at certain places, enter the desired depth, *in pencil*, in the last column, opposite the stakes marking those places. Then add together this depth and the corresponding surface measurement in the column next preceding, and enter the sum, *in pencil*, in the fifth column, as the depth from the datum-line to the desired position of the drain. (In the example in hand, these points are at Nos. 3, 7, 10, 11, and 17.)

7th. Subtract the second amount in the fifth column from the first amount for the total fall between the two points—in the example, "3" from "Silt-Basin." Divide this total fall, (in feet and hundredths,) by one hundredth of the total number of feet between them. The result will be the rate of fall per 100 feet, and this should be entered, in the third column, opposite each of the intermediate distances between the points.

Example: Depth of the Drain at the Silt-Basin..........20.45 feet.
" " " " " Stake No. 3...........16.98 "

Difference................................ 3.47 "
Distance between the two..................186.— "

```
1.8 6) 3.4 7 (1.8 6 5 or 1.8 7
       1 8 6
       -----
       1 6 1 0
       1 4 8 8
       -------
         1 2 2 0
         1 1 1 6
         -------
           1 0 4 0
             9 3 0
           -------
             1 1 0
```

8th. Multiply the numbers of the second column by those of the third and divide the product by 100. The result will be the amount of fall between the stakes, (fourth column.)—Example: 1.87 × 82=153÷100=1.53.

9th. Subtract the first number of the fourth column from the first number of the fifth column, (on the line above it,) and place the remainder on the next line of the fifth column.—Example: 20.70−1.64=19.06.

Then, from this new amount, subtract the second number of the fourth column, for the next number of the fifth, and so on, until, in place of the entry in pencil, (Stake 3,) we place the exact result of the computation.

Proceed in like manner with the next interval,—3 to 7.

10th. Subtract the numbers in the sixth column from those in the fifth, and the remainders will be the depths to be entered in the last.

Under the head of "Remarks," note any peculiarity of the drain which may require attention in the field.

The main lines *A*, *D*, and *E*, and the drain *B*, should next be graded on the plan set forth for *C*, and their laterals, all of which have considerable fall, and being all so steep as not to require silt-basins at any point,—can, by a very simple application of the foregoing principles, be adjusted at the proper depths. In grading the stone and tile drain, (*H*, *I*,) it is only necessary to adopt the depth of the last stakes of the laterals, with which it is connected, as it is immaterial in which direction the water flows. The ends of this drain,—from *H* to the head of the drain *C* 10, and from *I* to the head of *C* 17,—should, of course, have a decided fall toward the drains.

The laterals which are placed at intervals of 20 feet, over the underground rock on the east side of the field, should be continued at a depth of about 3 feet for nearly their whole length, dropping in a distance of 8 or 10 feet at their lower ends to the top of the tile of the main. The intervals between the lower ends of *C 7c*, *C 7d*, and *C 7e*, being considerably more than 20 feet, the drains may be gradually deepened, throughout their whole length from 3 feet at the upper ends to the depth of the top of the main at the lower ends.

The main drains *F* and *G*, being laid in flat land, their

outlets being fixed at a depth of 3.50, (the floor of the main outlet,) and it being necessary to have them as deep as possible throughout their entire length, should be graded with great care on the least admissible fall. This, in ordinary agricultural drainage, may be fixed at .25, or 3 inches, per 100 feet. Their laterals should commence with the top of their $\frac{1}{4}$ tile even with the top of the $2\frac{1}{2}$ collar of the main,—or .15 higher than the grade of the main,—and rise, at a uniform inclination of .25, to the upper end.

Having now computed the depth at which the tile is to lie, at each stake, and entered it on the map, we are ready to mark these depths on their respective stakes in the field, when the preliminary engineering of the work will be completed.

It has been deemed advisable in this chapter to consider the smallest details of the work of the draining engineer. Those who intend to drain in the best manner will find such details important. Those who propose to do their work less thoroughly, may still be guided by the principles on which they are based. Any person who will take the pains to mature the plans of his work as closely as has been here recommended, will as a consequence commence his operations in the field much more understandingly. The advantage of having everything decided beforehand,—so that the workmen need not be delayed for want of sufficient directions, and of making, on the map, such alterations as would have appeared necessary in the field, thus saving the cost of cutting ditches in the wrong places, will well repay the work of the evenings of a whole winter.

CHAPTER IV.

HOW TO MAKE THE DRAINS.

Knowing, now, precisely what is to be done; having the lines all staked out, and the stakes so marked as to be clearly designated; knowing the precise depth at which the drain is to be laid, at every point; having the requisite tiles on the ground, and thoroughly inspected, the operator is prepared to commence actual work.

He should determine how many men he will employ, and what tools they will require to work to advantage. It may be best that the work be done by two or three men, or it may be advisable to employ as many as can work without interfering with each other. In most cases,—especially where there is much water to contend with,—the latter course will be the most economical, as the ditches will not be so liable to be injured by the softening of their bottoms, and the caving in of their sides.

The Tools Required are a subsoil plow, two garden lines, spades, shovels, and picks; narrow finishing spades, a finishing scoop, a tile pick, a scraper for filling the ditches, a heavy wooden maul for compacting the bottom filling, half a dozen boning-rods, a measuring rod, and a plumb rod. These should all be on hand at the outset, so that no delay in the work may result from the want of them.

Writers on drainage, almost without exception, recommend the use of elaborate sets of tools which are intended

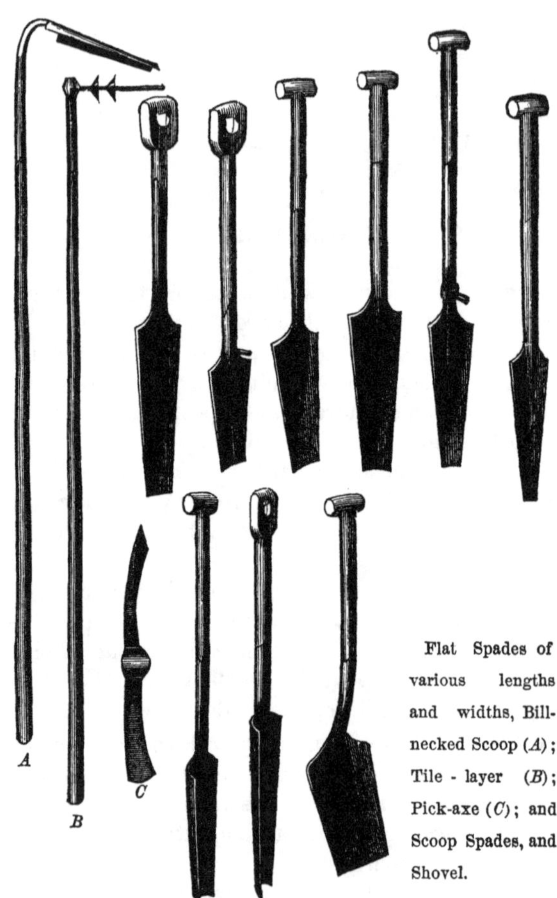

Flat Spades of various lengths and widths, Bill-necked Scoop (*A*); Tile-layer (*B*); Pick-axe (*C*); and Scoop Spades, and Shovel.

Fig. 22.—SET OF TOOLS.

for cutting very narrow ditches,—only wide enough at the bottom to admit the tile, and not allowing the workmen to stand in the bottom of the ditch. A set of these tools is shown in Fig. 22.

Possibly there may be soils in which these implements, in the hands of men skilled in their use, could be employed with economy, but they are very rare, and it is not believed to be possible, under any circumstances, to regulate the bottom of the ditch so accurately as is advisable, unless the workman can stand directly upon it, cutting it more smoothly than he could if the point of his tool were a foot or more below the level on which he stands.

On this subject, Mr. J. Bailey Denton, one of the first draining engineers of Great Britain, in a letter to Judge French, says:

"As to tools, it is the same with them as it is with the "art of draining itself,—too much rule and too much draw-"ing upon paper; all very right to begin with, but very "prejudicial to progress. I employ, as engineer to the "General Land Drainage Company, and on my private "account, during the drainage season, as many as 2,000 "men, and it is an actual fact, that not one of them uses "the set of tools figured in print. I have frequently pur-"chased a number of sets of the Birmingham tools, and "sent them down on extensive works. The laborers would "purchase a few of the smaller tools, such as Nos. 290, "291, and 301, figured in Morton's excellent Cyclopædia "of Agriculture, and would try them, and then order "others of the country blacksmith, differing in several "respects; less weighty and much less costly, and more-"over, much better as working tools. All I require of the "cutters, is, that the bottom of the drain should be evenly "cut, to fit the size of the pipe. The rest of the work "takes care of itself; for a good workman will economize "his labor for his own sake, by moving as little earth as "practicable; thus, for instance, a first-class cutter, in "clays, will get down 4 feet with a 12-inch opening, *ordi-* "*narily;* if he wishes to *show off*, he will sacrifice his "own comfort to appearance, and will do it with a 10-inch "opening."

In the Central Park work, sets of these tools were procured, at considerable expense, and every effort was made to compel the men to use them, but it was soon found that, even in the easiest digging, there was a real economy in using, for the first 3 feet of the ditch, the common spade, pick, and shovel,—finishing the bottoms with the narrow spade and scoop hereafter described, and it is probable that the experience of that work will be sustained by that of the country at large.

Marking the Lines.—To lay a drain directly under the position of its stakes, would require that enough earth be left at each point to hold the stake, and that the ditch be tunneled under it. This is expensive and unnecessary. It is better to dig the ditches at one side of the lines of stakes, far enough away for the earth to hold them firmly in their places, but near enough to allow measurements to be taken from the grade pegs. If the ditch be placed always to the right, or always to the left, of the line, and at a uniform distance, the general plan will remain the same, and the lines will be near enough to those marked on the map to be easily found at any fúture time. In fact, if it be known that the line of tiles is two feet to the right of the position indicated, it will ȯnly be necessary, at any time, should it be desired to open an old drain, to measure two feet to the right of the surveyed position to strike the line at once.

In soils of ordinary tenacity, ditches 4 feet deep need not be more than twenty (20) inches wide at the surface, and four (4) inches wide at the bottom. This will allow, in each side, a slope of eight (8) inches, which is sufficient except in very loose soils, and even these may be braced up, if inclined to cave in. There are cases where the soil contains so much running sand, and is so saturated with water, that no precautions will avail to keep up the banks. Ditches in such ground will sometimes fall in, until the excavation reaches a width of 8 or 10 feet. Such instances, however, are very rare, and must be treated as the occasion suggests.

One of the garden lines should be set at a distance of about 6 inches from the row of stakes, and the other at a further distance of 20 inches. If the land is in grass, the position of these lines may be marked with a spade, and they may be removed at once; but, if it is arable land, it will be best to leave the lines in position until the ditch is excavated to a sufficient depth to mark it clearly. Indeed, it will be well to at once remove all of the sod and surface soil, say to a depth of 6 inches, (throwing this on the same side with the stakes, and back of them.) The whole force can be profitably employed in this work, until all of the ditches to be dug are scored to this depth over the entire tract to be drained, except in swamps which are still too wet for this work.

Water Courses.—The brooks which carry the water from the springs should be "jumped" in marking out the lines, as it is desirable that their water be kept in separate channels, so far as possible, until the tiles are ready to receive it, as, if allowed to run in the open ditches, it would undermine the banks and keep the bottom too soft for sound work.

With this object, commence at the southern boundary of our example tract, 10 or 15 feet east of the point of outlet, and drive a straight, temporary, shallow ditch to a point a little west of the intersection of the main line *D* with its first lateral; then carry it in a northwesterly direction, crossing *C* midway between the silt-basin and stake *C* 1, and thence into the present line of the brook, turning all of the water into the ditch. A branch of this ditch may be run up between the lines *F* and *G* to receive the water from the spring which lies in that direction. This arrangement will keep the water out of the way until the drains are ready to take it.

The Outlet.—The water being all discharged through the new temporary ditch, the old brook, beyond the boundary, should be cleared out to the final level (3.75,) and an excavation made, just within the boundary, sufficient to receive the masonry which is to protect the outlet. A good form of outlet is shown in Fig. 23. It may

Fig. 23.—OUTLET, SECURED WITH MASONRY AND GRATING.

be cheaply made by any farmer, especially if he have good stone at hand;—if not, brick may be used, laid on a solid foundation of stout planks, which, (being protected from the air and always saturated with water,) will last a very long time.

If made of stone, a solid floor, at least 2 feet square, should be placed at, or below, the level of the brook. If this consist of a single stone, it will be better than if of several smaller pieces. On this, place another layer extending the whole width of the first, but reaching only from its inner edge to its center line, so as to leave a foot in width of the bottom stone to receive the fall of the water. This second layer should reach exactly the grade of the outlet (3.50) or a height of 3 inches from the brook level. On the floor thus made, there should be laid the tiles which are to constitute the outlets of the several drains; *i. e.*, one $3\frac{1}{2}$-inch tile for the line from the silt-basin, two $1\frac{1}{4}$-inch for the lines *F* and *G*, and one $2\frac{1}{4}$-inch for the main line *E*. These tiles should lie close to each other and be firmly cemented together, so that no water can pass outside of them, and a rubble-work of stone may with advantage be carried up a foot above them. Stone work, which may be rough and uncemented, but should always be solid, may then be built up at the sides, and covered with a secure coping of stone. A floor and sloping sides of stone work, jointed with the previously described work, and well cemented, or laid in strong clay or mortar, may, with benefit, be carried a few feet beyond the outlet. This will effectually prevent the undermining of the structure. After the entire drainage of the field is finished, the earth above these sloping sides, and that back of the coping, should be neatly sloped, and protected by sods. An iron grating, fine enough to prevent the entrance of vermin, placed in front of the tile, at a little distance from them,—and secured by a flat stone set on edge and hollowed out, so as merely to allow the water to flow freely from the drains,—the stone being cemented in its place so as to allow no water to pass under it,—will give a substantial and permanent finish to the structure.

An outlet finished in this way, at an extra cost of a few dollars, will be most satisfactory, as a lasting means of securing the weakest and most important part of the system of drains. When no precaution of this sort is taken, the water frequently forces a passage under the tile for some distance up the drains, undermining and displacing them, and so softening the bottom that it will be difficult, in making repairs, to secure a solid foundation for the work.

Usually, repairs of this sort, aside from the annoyance attending them, will cost more than the amount required to make the permanent outlet described above. As well constructed outlets are necessarily rather expensive, as much of the land as possible should be drained to each one that it is necessary to make, by laying main lines which will collect all of the water which can be brought to it.

The Main Silt-Basin.—The silt-baisn, at which the drains are collected, may best be built before any drains are brought to it, and the work may proceed simultaneously with that at the outlet. It should be so placed that its center will lie exactly under the stake which marks its position, because it will constitute one of the leading landmarks for the survey of the drains.*

Before removing the stake and grade stake, mark their position by four stakes, set at a distance from it of 4 or 5 feet, in such positions that two lines, drawn from those which are opposite to each other, will intersect at the point indicated; and place near one of them a grade stake, driven to the exact level of the one to be removed. This being done, dig a well, 4 feet in diameter, to a depth of 2½ feet below the grade of the outlet drain, (in the example under consideration this would be 5 feet below the grade stake.) If much water collects in the hole, widen it, in the direction of the outlet drain, sufficiently to give room for baling out the water. Now build, in this well, a structure 2 feet in interior diameter, such as is shown in Fig. 24, having its bottom 2 feet, in the clear, below the grade of the outlet, and carry its wall a little higher than the general surface of the ground. At the proper height insert, in the brick work, the necessary for tiles all incoming and outgoing drains; in this case, a 3½-inch tile for

* The drains, which are removed a little to one side of the lines of stakes, may be turned toward the basin from a distance of 3 or 4 feet.

the outlet, $2\frac{1}{4}$-inch for the mains *A* and *C*, and $1\frac{1}{4}$-inch for *B* and *D*.

This basin being finished and covered with a flat stone or other suitable material, connect it with the outlet by an open ditch, unless the bottom of the ditch, when laid open to the proper depth, be found to be of muck or quicksand. In such case, it will be best to lay the tile at once, and cover it in for the whole distance, as, on a soft bottom, it would be difficult to lay it well when the full drainage of the field is flowing through the ditch. The tiles should be laid with all care, on a perfectly regulated fall,—using strips of board under them if the bottom is shaky or soft,—as on this line depends the success of all the drains above it, which might be rendered useless by a single badly laid tile at this point, or by any other cause of obstruction to the flow.

Fig. 24.—SILT-BASIN, BUILT TO THE SURFACE.

While the work is progressing in the field above, there will be a great deal of muddy water and some sticks, grass, and other rubbish, running from the ditches above the basin, and care must be taken to prevent this drain from becoming choked. A piece of wire cloth, or basket work, placed over the outlet in the basin, will keep out the coarser matters, and the mud which would accumulate in the tile may be removed by occasional flushing. This is done by crowding a tuft of grass,—or a bit of sod,—into the lower end of the tile (at the outlet,) securing it there until the water rises in the basin, and then removing it. The rush of water will be sufficient to wash the tile clean.

This plan is not without objections, and, as a rule, it is never well to lay any tiles at the lower end of a drain until all above it is finished; but when a considerable outlet must be secured through soft land, which is inclined to cave in, and to get soft at the bottom, it will save labor to secure the tile in place before much water reaches it, even though it require a daily flushing to keep it clean.

Opening the Ditches.—Thus far it has been sought to secure a permanent outlet, and to connect it by a secure channel, with the silt-basin, which is to collect the water of the different series of drains. The next step is to lay open the ditches for these. It will be best to commence with the main line *A* and its laterals, as they will take most of the water which now flows through the open brook, and prevent its interference with the rest of the work.

The first work is the opening of the ditches to a depth of about 3 feet, which may be best done with the common spade, pick, and shovel, except that in ground which is tolerably free from stones, a subsoil plow will often take the place of the pick, with much saving of labor. It *may* be drawn by oxen working in a long yoke, which will allow them to walk one on each side of the ditch, but this is dangerous, as they are liable to disturb the stakes, (especially the grade stakes,) and to break down the edges of the ditches. The best plan is to use a small subsoil plow, drawn by a single horse, or strong mule, trained to walk in the ditch. The beast will soon learn to accommodate himself to his narrow quarters, and will work easily in a ditch 2½ feet deep, having a width of less than a foot at the bottom; of course there must be a way provided for him to come out at each end. Deeper than this there is no economy in using horse power, and even for this depth it will be necessary to use a plow having only one stilt.

Before the main line is cut into the open brook, this should be furnished with a wooden trough, which will carry the water across it, so that the ditch shall receive only the filtration from the ground. Those laterals west of the main line, which are crossed by the brook, had better not be opened at present,—not until the water of the spring is admitted to and removed by the drain.

The other laterals and the whole of the main line, having been cut to a depth of 3 feet, take a finishing spade, (Fig. 25,) which is only 4 inches wide at its point, and dig to within 2 or 3 inches of the depth marked on the stakes, making the bottom tolerably smooth, with the aid of the finishing scoop, (Fig. 26,) and giving it as regular an inclination as can be obtained by the eye alone.

Fig. 25.—FINISHING SPADE.

Often, large stones, which would cost much labor to remove, will be encountered in the digging. If these lie from 6 inches to a foot above the final grade, and are not too large, it will be easier to tunnel under them than to take them out, or to go around them; but, if they are very large, or lie close to the bottom, (or in the bottom,) the latter course will be necessary.

Fig. 26.—FINISHING SCOOP.

If the ground is "rotten," and the banks of the ditches incline to cave in, as is often the case in passing wet places, the earth which is thrown out in digging must be thrown back sufficiently far from the edge to prevent its weight from increasing the tendency; and the sides of the ditch may be supported by bits of board braced apart as is shown in Fig. 27.

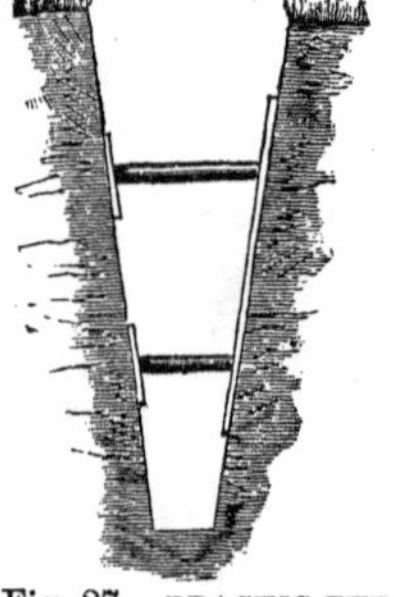
Fig 27.—BRACING THE SIDES IN SOFT LAND.

The manner of opening the ditches, which is described above, for the main *A* and its laterals, will apply to the drains of the whole field and to all similar work.

Grading the Bottoms.—The next step in the work is to grade the bottoms of the ditches, so as to afford a bed for the tiles on the exact lines which are indicated by the figures marked on the different stakes.

The manner in which this is to be done may be illustrated by describing the work required for the line from *C* 10 to *C* 17, (Fig. 20,) after it has been opened, as described above, to within 2 or 3 inches of the final depth.

A measuring rod, or square, such as is shown in Fig. 28,* is set at *C* 10, so that the lower side of its arm is at the mark 4.59 on the staff, (or at a little less than 4.6 if it is divided only into feet and tenths,) and is held upright in the ditch, with its arm directly over the grade stake. The earth below it is removed, little by little, until it will touch the top of the stake and the bottom of the ditch at the

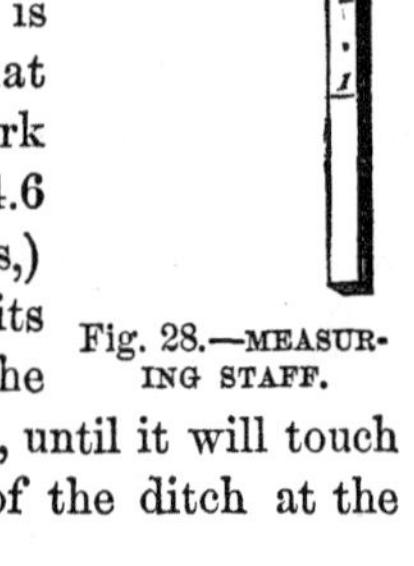

Fig. 28.—MEASURING STAFF.

* The foot of the measuring rod should be shod with iron to prevent its being worn to less than the proper length.

same time. If the ground is soft, it should be cut out until a flat stone, a block of wood, or a piece of tile, or of brick, sunk in the bottom, will have its surface at the exact point of measurement. This point is the bottom of the ditch on which the collar of the tile is to lie at that stake. In the same manner the depth is fixed at *C* 11 (4.19,) and *C* 12 (4.41,) as the rate of fall changes at each of these points, and at *C* 15 (3.89,) and *C* 17 (4.17,) because (although the fall is uniform from *C* 12 to *C* 17,) the distance is too great for accurate sighting.

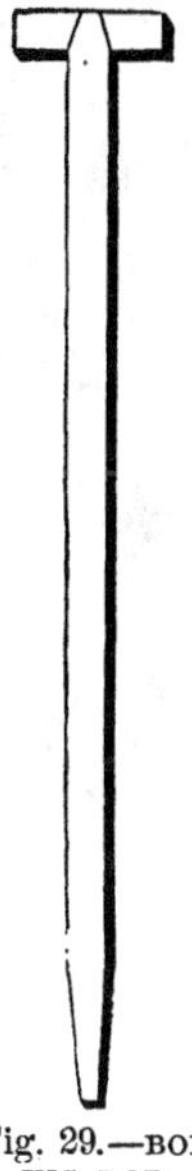

Having provided *boning-rods*, which are strips of board 7 feet long, having horizontal cross pieces at their upper ends, (see Fig. 29,) set these perpendicularly on the spots which have been found by measurement to be at the correct depth opposite stakes 10, 11, 12, 15, and 17, and fasten each in its place by wedging it between two strips of board laid across the ditch, so as to clasp it, securing these in their places by laying stones or earth upon their ends.

Fig. 29.—BONING ROD.

As these boning-rods are all exactly 7 feet long, of course, a line sighted across their tops will be exactly 7 feet higher, at all points, than the required grade of the ditch directly beneath it, and if a plumb rod, (similar to the boning-rod, but provided with a line and plummet,) be set perpendicularly on any point of the bottom of the drain, the relation of its cross piece to the line of sight across the tops of the boning-rods will show whether the bottom of the ditch at that point is too high, or too low, or just right. The manner of sighting over two boning-rods and an intermediate plumb-rod, is shown in Fig. 31.

Three persons are required to finish the bottom of the

ditch; one to sight across the tops of the boning-rods, one to hold the plumb-rod at different points as the finishing progresses, and one in the ditch, (see Fig. 30,) provided with the finishing spade and scoop,—and, in hard ground, with a pick,—to cut down or fill up as the first man calls

Fig. 30.—POSITION OF WORKMAN AND USE OF FINISHING SCOOP.

"too high," or "too low." An inch or two of filling may be beaten sufficiently hard with the back of the scoop, but if several inches should be required, it should be well

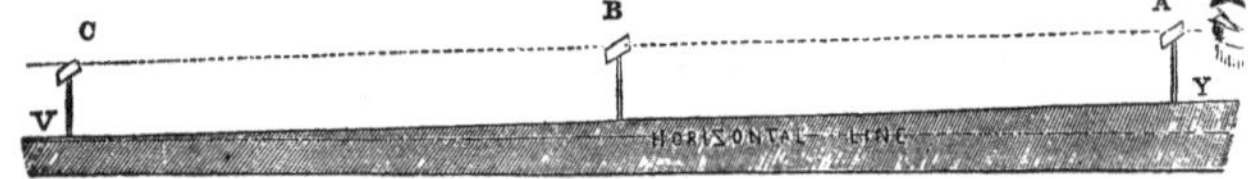

Fig. 31.—SIGHTING BY THE BONING-RODS.

rammed with the top of a pick, or other suitable instrument, as any subsequent settling would disarrange the fall.

As the lateral drains are to be laid first, they should be the first graded, and as they are arranged to discharge into the tops of the mains, their water will still flow off, although the main ditches are not yet reduced to their final depth. After the laterals are laid and filled in, the main should be graded, commencing at the upper end; the tiles being laid and covered as fast as the bottom is made ready, so that it may not be disturbed by the water of which the main carries so much more than the laterals.

Tile-Laying.—Gisborne says: "It would be scarcely "more absurd to set a common blacksmith to eye needles "than to employ a common laborer to lay pipes and col- "lars." The work comes under the head of *skilled labor*, and, while no very great exercise of judgment is required in its performance, the little that is required is imperatively necessary, and the details of the work should be deftly done. The whole previous outlay,—the survey and staking of the field, the purchase of the tiles, the digging and grading of the ditches—has been undertaken that we may make the conduit of earthenware pipes which is now to be laid, and the whole may be rendered useless by a want of care and completeness in the performance of this chief operation. This subject, (in connection with that of finishing the bottoms of the ditches,) is very clearly treated in Mr. Hoskyns' charming essay,* as follows:

"It was urged by Mr. Brunel, as a justification for more 'attention and expense in the laying of the rails of the "Great Western, than had been ever thought of upon "previously constructed lines, that all the embankments "and cuttings, and earthworks and stations, and law and "parliamentary expenses—in fact, the whole of the out- "lay encountered in the formation of a railway, had for its "main and ultimate object *a perfectly smooth and level* "*line of rail;* that to turn stingy at this point, just when "you had arrived at the great ultimatum of the whole "proceedings, viz: the iron wheel-track, was a sort of "saving which evinced a want of true preception of the "great object of all the labor that had preceded it. It

* "Talpa, or the Chronicles of a Clay Farm."

"may seem curious to our experiences, in these days, that "such a doctrine could ever have needed to be enforced "by argument; yet no one will deem it wonderful who "has personally witnessed the unaccountable and ever new "difficulty of getting proper attention paid to the leveling "of the bottom of a drain, and the laying of the tiles in "that continuous line, where one single depression or ir- "regularity, by collecting the water at that spot, year "after year, tends toward the eventual stoppage of the "whole drain, through two distinct causes, the softening "of the foundation underneath the sole, or tile flange, and "the deposit of soil inside the tile from the water collected "at the spot, and standing there after the rest had run off. "Every depression, however slight, is constantly doing "this mischief in every drain where the fall is but trifling; "and if to the two consequences above mentioned, we "may add the decomposition of the tile itself by the "action of water long stagnant within it, we may deduce "that every tile-drain laid with these imperfections in "the finishing of the bottom, has a tendency toward "obliteration, out of all reasonable proportion with "that of a well-burnt tile laid on a perfectly even inclina- "tion, which, humanly speaking, may be called a perma- "nent thing. An open ditch cut by the most skillful "workman, in the summer, affords the best illustration of "this underground mischief. Nothing can look smoother "and more even than the bottom, until that uncompromis- "ing test of accurate levels, the water, makes its appear- "ance: all on a sudden the whole scene is changed, the "eye-accredited level vanishes as if some earthquake had "taken place: here, there is a gravelly *scour*, along which "the stream rushes in a thousand little angry-looking rip- "ples; there, it hangs and looks as dull and heavy as if it "had given up running at all, as a useless waste of energy; "in another place, a few dead leaves or sticks, or a morsel "of soil broken from the side, dams back the water for a "considerable distance, occasioning a deposit of soil along "the whole reach, greater in proportion to the quantity "and the muddiness of the water detained. All this shows "the paramount importance of perfect evenness in the "bed on which the tiles are laid. *The worst laid tile is "the measure of the goodness and permanence of the "whole drain*, just as the weakest link of a chain is the "measure of its strength."

The simple laying of the smaller sizes of pipes and collars in the lateral drains, is an easy matter. It requires care and precision in placing the collar equally under the end of each pipe, (having the joint at the middle of the collar,) in having the ends of the pipes actually touch each other within the collars, and in brushing away any loose dirt which may have fallen on the spot on which the collar is to rest. The connection of the laterals with the mains, the laying of the larger sizes of tiles so as to form a close joint, the wedging of these larger tiles firmly into their places, and the trimming which is necessary in going around sharp curves, and in putting in the shorter pieces which are needed to fill out the exact length of the drain, demand more skill and judgment than are often found in the common ditcher. Still, any clever workman, who has a careful habit, may easily be taught all that is necessary; and until he is thoroughly taught,—and not only knows how to do the work well, but, also, understands the importance of doing it well,—the proprietor should carefully watch the laying of every piece.

Never have tiles laid by the rod, but always by the day. "The more haste, the less speed," is a maxim which applies especially to tile-laying.

If the proprietor or the engineer does not overlook the laying of each tile as it is done, and probably he will not, he should carefully inspect every piece before it is covered. It is well to walk along the ditches and touch each tile with the end of a light rod, in such a way as to see whether it is firm enough in its position not to be displaced by the earth which will fall upon it in filling the ditches.

Preparatory to laying, the tiles should be placed along one side of the ditch, near enough to be easily reached by a man standing in it. When collars are to be used, one of these should be slipped over one end of each tile. The workman stands in the ditch, with his face toward its upper end. The first tile is laid with a collar on its lower end, and the collar is drawn one-half of its length forward, so as to receive the end of the next tile. The upper end of the first tile is closed with a stone, or a bit of broken tile placed firmly against it. The next tile has its nose placed into the projecting half of the collar of the first one, and its own collar is drawn forward to receive the end of the third, and thus to the end of the drain, the workman walking backward as the work progresses. By and by, when he comes to connect the lateral with the main, he may find that a short piece of tile is needed to complete the length; this should not be placed next to the tile of the main, where it is raised above the bottom of the ditch, but two or three lengths back, leaving the connection with the main to be made with a tile of full length. If the piece to be inserted is only two or three inches long, it may be omitted, and the space covered by using a whole 2½-inch tile in place of the collar. In turning corners or sharp curves, the end of the tile may be chipped off, so as to be a little thinner on one side, which will allow it to be turned at a greater angle in the collar.

If the drain turns a right angle, it will be better to dig out the bottom of the ditch to a depth of about eight inches, and to set a 6-inch tile on end in the hole, perforating its sides, so as to admit the ends of the pipes at the proper level. This 6-inch tile, (which acts as a small silt-basin,) should stand on a board or on a flat stone, and its top should be covered with a stone or with a couple of bricks. Wood will last almost forever below the level of the drain, where it will always be saturated with water, but in the drier earth above the tile, it is much more liable to decay.

The trimming and perforating of the tile is done with a "tile-pick," (Fig. 32,) the hatchet end, tolerably sharp, being used for the trimming, and the point, for making the holes. This is done by striking lightly around the circumference of the hole until the center piece falls in, or can be easily knocked in. If the hole is irregular, and does not fit the tile nicely, the open space should be covered with bits of broken tile, to keep the earth out.

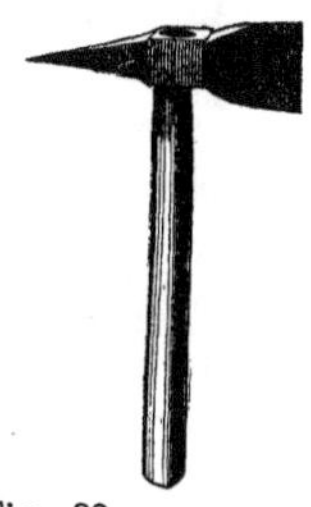

Fig. 32.—PICK FOR DRESSING AND PERFORATING TILE.

As fast as the laterals are laid and inspected, they should be filled in to the depth of at least a foot, to protect the tiles from being broken by the falling of stones or lumps of earth from the top, and from being displaced by water flowing in the ditch. Two or three feet of the lower end may be left uncovered until the connection with the main is finished.

In the main drains, when the tiles are of the size with which collars are used, the laying is done in the same manner. If it is necessary to use 3½-inch tiles, or any larger size, much more care must be given to the closing of the joints. All tiles, in manufacture, dry more rapidly at the top, which is more exposed to the air, than at the bottom, and they are, therefore, contracted and made shorter at the top. This difference is most apparent in the larger sizes. The large *round* tiles, which can be laid on any side, can easily be made to form a close joint, and they should be secured in their proper position by stones or lumps of earth, wedged in between them and the sides of the ditch. The sole tiles must lie with the shortest sides up, and, usually, the space between two tiles, at the top, will be from one-quarter to one-half of an inch. To remedy this defect, and form a joint which may be protected against the entrance of earth, the bottom should be trimmed off,.so as to allow the tops to come closer together. Any opening, of less than a quarter of an inch, can be satisfactorily covered,—more than that should not be allowed. In turning corners, or in passing around curves, with large tiles, their ends must be beveled off with the pick, so as to fit nicely in this position.

The best covering for the joints of tiles which are laid without collars, is a scrap of tin, bent so as to fit their shape,—scraps of leather, or bits of strong wood shavings, answer a very good purpose, though both of these latter require to be held in place by putting a little earth over their ends as soon as laid on the tile. *Very small* grass ropes drawn over the joints, (the ends being held down with stones or earth,) form a satisfactory covering, but care should be taken that they be not too thick. A small handful of wood shavings, thrown over the joints, also answers a good purpose. Care, however, should always be taken, in using any material which will decay readily, to have no more than is necessary to keep the earth out, lest, in its decay, it furnish material to be carried into the tile and obstruct the flow. This precaution becomes less necessary in the case of drains which always carry considerable streams of water, but if they are at times sluggish in their flow, too much care cannot be given to keep them free of all possible causes of obstruction. As nothing is gained by increasing the quantity of loose covering beyond what is needed to close the joints, and as such covering is only procured with some trouble, there is no reason for its extravagant use.

There seems to remain in the minds of many writers on drainage a glimmering of the old fallacy that underdrains, like open drains, receive their water from above, and it is too commonly recommended that porous substances be placed above the tile. If, as is universally conceded, the water rises into the tile from below, this is unnecessary. The practice of covering the joints, and even covering the whole tile, (often to the depth of a foot,) with tan-bark, turf, coarse gravel, etc., is in no wise to be commended; and, while the objections to it are not necessarily very grave in all cases, it always introduces an element of insecurity, and it is a waste of money, if nothing worse.

The tile layer need not concern himself with the question of affording entrance room for the water. Let him, so far as the rude materials at hand will allow, make the joints perfectly tight, and when the water comes, it will find ample flaws in his work, and he will have been a good workman if it do not find room to flow in a current, carrying particles of dirt with it.

In ditches in which water is running at the time of laying the tiles, the process should follow closely after the grading, and the stream may even be dammed back, section after section, (a plugged tile being placed under the dam, to be afterwards replaced by a free one,) and graded, laid and covered before the water breaks in. There is one satisfaction in this kind of work,—that, while it is difficult to lay the drain so thoroughly well as in a dry ditch, the amount of water is sufficient to overcome any slight tendency to obstruction.

Connections.—As has been before stated, lateral drains should always enter at the top of the main. Even in the most shallow work, the slightly decreased depth of the lateral, which this arrangement requires, is well compensated for by the free outlet which it secures.

After the tile of the main, which is to receive a side drain, has been fitted to its place, and the point of junction marked, it should be taken up and perforated; then the end of the tile of the lateral should be so trimmed as to fit the hole as accurately as may be, the large tile replaced in its position, and the small one laid on it,—reaching over to the floor of the lateral ditch. Then connect it with the lateral as previously laid, fill up solidly the space under the tile which reaches over to the top of the main, (so that it cannot become disturbed in filling,) and lay bits of tile, or other suitable covering, around the connecting joint.*

When the main drain is laid with collars, it should be so arranged that, by substituting a full tile in the place of the collar,—leaving, within it, a space between the smaller pipes,—a connection can be made with this larger tile, as is represented in Figures 33 and 34.

Fig. 33.—LATERAL DRAIN ENTERING AT TOP.

Silt-Basins should be used at all points where a drain, after running for any considerable distance at a certain rate of fall, changes to a less rapid fall,—unless, indeed, the diminished fall be still sufficiently great for the removal of silty matters, (say two feet or more in a hundred). They may be made in any manner which will secure a stoppage of the direct current, and afford room below the floor of the tile for the deposit of the silt which the water has carried in suspension; and they may be of any suitable material;—even a sound flour barrel will serve a pretty good

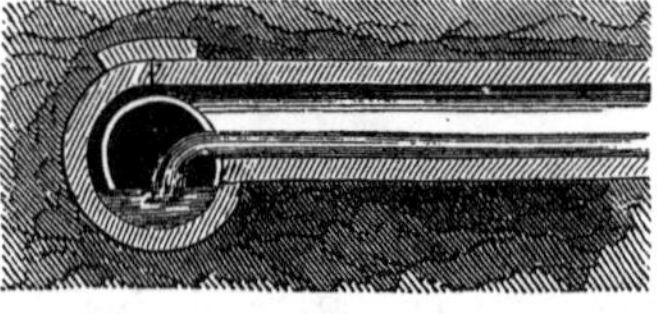

Fig. 34.—SECTIONAL VIEW OF JOINT.

*When chips of tile, or similar matters, are used to cover openings in the tile-work, it is well to cover them at once with a mortar made of wet clay, which will keep them in place until the ditches are filled.

purpose for many years. The most complete form of basin is that represented in Figure 24.

Fig. 35.—SQUARE BRICK SILT-BASIN.

When the object is only to afford room for the collection of the silt of a considerable length of drain, and it is not thought worth while to keep open a communication with the surface, for purposes of inspection, a square box of brick work, (Fig. 35,) having a depth of one and a half or two feet below the floor of the drain,—tiles for the drains being built in the walls, and the top covered with a broad stone,—will answer very well.

A good sort of basin, to reach to the surface of the ground, may be made of large, vitrified drain pipes,—such as are used for town sewerage,—having a diameter of from six to twelve inches, according to the requirements of the work. This basin is shown in Figure 36.

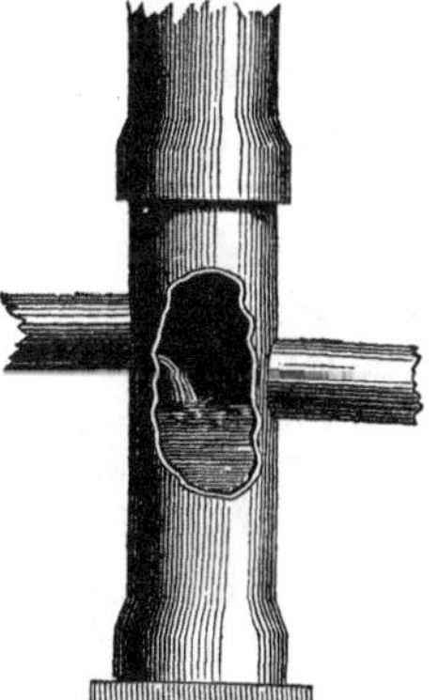

Fig. 36.—SILT-BASIN OF VITRIFIED PIPE.

Figure 37 represents a basin made of a 6-inch tile,—similar to that described on page 130, for turning a short corner. A larger basin of the same size, cheaper than if built of brick, may be made by using a large vitrified drain pipe in the place of the one shown in the cut. These vitrified pipes may be perforated in the manner described for the common tile.

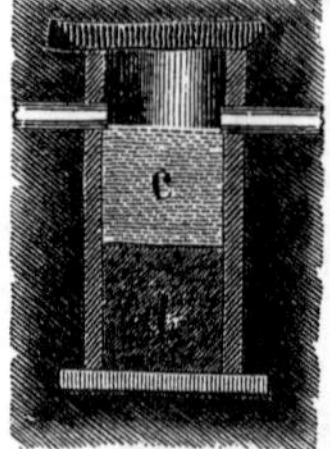
Fig. 37.—TILE SILT-BASIN.

In laying the main line *C*, (Fig. 21,) an underground basin of brick work, (Fig. 35,) or its equivalent, should be placed at stake 7, because at that point the water, which has been flowing on an inclination of 1.09, 2.00 and 2.83 per 100, continues its course over the much less fall of only 0.56 per 100.

If, among the tiles which have passed the inspection, there are some which, from over burning, are smaller than the average, they should be laid at the upper ends of the laterals. The cardinal rule of the tile layer should be *never to have a single tile in the finished drain of smaller size, of more irregular shape, or less perfectly laid, than any tile above it.* If there is to be any difference in the quality of the drain, at different points, let it grow better as it approaches the outlet and has a greater length above depending upon its action.

Covering the Tiles, and Filling-in the Ditches.—The best material for covering the tiles is that which will the most completely surround them, so as to hold them in their places; will be the least likely to have passages for the flow of *streams* of water into the joints, and will afford the least silt to obstruct the drain. Clay is the best of all available materials, because it is of the most uniform character throughout its mass, and may be most perfectly compacted around the tiles. As has been before stated, all matters which are subject to decay are objectionable, because they will furnish fine matters to enter the joints, and by their decrease of bulk, may leave openings in the earth through which streams of muddy water may find their way into the tiles. Gravel is bad, and will remain bad until its spaces are filled with fine dirt deposited by water, which, leaving only a part of its impurities here, carries the rest into the drain. A gravelly loam, free from roots or other organic matter, if it is strong enough to be worked into a ball when wet, will answer a very good purpose.

Ordinarily, the earth which was thrown out from the bottom of the ditch, and which now lies at the top of the dirt heap, is the best to be returned about the tiles, being first freed from any stones it may contain which are large enough to break or disturb the tiles in falling on to them.

If the bottom of the ditch consists of quicksand or other silty matters, clay or some other suitable earth should be sought in that which was excavated from a less depth, or should be brought from another place. A thin layer of this having been placed in the bottom of the ditch when grading, a slight covering of the same about the tiles will so encase them as to prevent the entrance of the more "slippy" soil.

The first covering of fine earth, free from stones and clods, should be sprinkled gently over the tiles, no full shovelfuls being thrown on to them until they are covered at least six inches deep. When the filling has reached a hight of from fifteen to twenty inches, the men may jump into the ditch and tramp it down evenly and regularly, not treading too hard in any one place at first. When thus lightly compacted about the tile, so that any further pressure cannot displace them, the filling should be repeatedly rammed, (the more the better,) by two men standing astride the ditch, facing each other, and working a maul, such as is shown in Figure 38, and which may weigh from 80 to 100 pounds.

Those to whom this recommendation is new, will, doubtless, think it unwise. The only reply to their objection must be that others who shared their opinion, have, by long observation and experience, been convinced of its correctness. They may practically convince themselves of the value of this sort of covering by a simple and inexpensive experiment: Take two large, water-tight hogsheads, bore through the side of each, a few inches from the bottom, a hole just large enough to admit a $1\frac{1}{4}$-inch tile; cover the bottom to the hight of the lower edge of the hole with strong, wet clay, beaten to a hard paste; on this, lay a line of pipes and collars,—the inner end sealed with putty, and the tile which passes through the hole so wedged about with putty, that no water could pass out between it and the outside of the hole. Cover the tile in one hogshead with loose gravel, and then fill it to the top with loose earth. Cover the tile in the other, twenty inches deep, with ordinary stiff clay, (not wet enough to *puddle*, but sufficiently moist to pack well,) and ram it thoroughly, so as to make sure that the tiles are completely clasped, and that there is no crack nor crevice through which water can trickle, and then fill this hogshead to the top with earth, of the same character with that used in the other case. These hogsheads should stand where the water of a small roof, (as that of a hog-pen,) may be led into them, by an arrangement which shall give an equal quantity to each;—this will give them rather more than the simple rain-fall, but will leave them exposed to the usual climatic changes of the season. A vessel, of a capacity of a quart or more, should be connected with each outlet, and covered from the dust,—

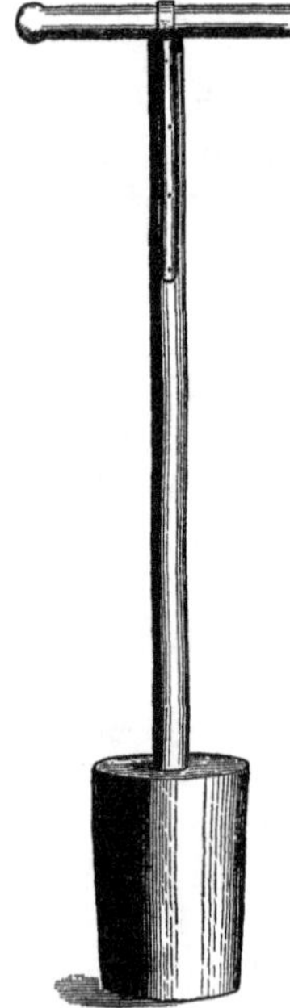
Fig. 38.—MAUL FOR RAMMING.

these will act as silt-basins. During the first few storms the water will flow off much more freely from the first barrel; but, little by little, the second one, as the water finds its way through the clay, and as the occasional drying and repeated filtration make it more porous, will increase in its flow until it will, by the end of the season, or, at latest, by the end of the second season, drain as well as the first, if, indeed, that be not by this time somewhat obstructed with silt. The amount of accumulation in the vessels at the outlet will show which process has best kept back the silt, and the character of the deposit will show which would most probably be carried off by the gentle flow of water in a nearly level drain.

It is no argument against this experiment that its results cannot be determined even in a year, for it is not pretended that drains laid in compact clay will dry land so completely during the first month as those which give more free access to the water; only that they will do so in a comparatively short time; and that, as drainage is a work for all time, (practically as lasting as the farm itself,) the importance of permanence and good working for long years to come, is out of all proportion to that of the temporary good results of one or two seasons, accompanied with doubtful durability.

It has been argued that *surface water* will be more readily removed by drains having porous filling. Even if this were true to any important degree,—which it is not,—it would be an argument against the plan, for the remedy would be worse than the disease. If the water flow from the surface down into the drain, it will not fail to carry dirt with it, and instead of the clear water, which alone should rise into the tiles from below, we should have a trickling flow from above, muddy with wasted manure and silty earth.

The remaining filling of the ditch is a matter of simple labor, and may be done in whatever way may be most economical under the circumstances of the work. If the amount to be filled is considerable, so that it is desirable to use horse-power, the best way will be to use a scraper, such as is represented in Figure 39, which is a strongly ironed plank, 6 feet long and 18 inches wide, sharp shod at one side, and supplied with handles at the other. It is propelled by means of the curved rods, which are attached to its under side by flexible joints. These rods are connected by a chain which has links large enough to

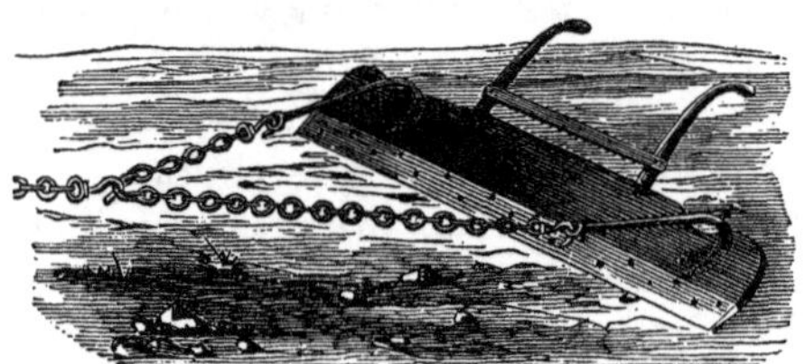

Fig. 39.—BOARD SCRAPER FOR FILLING DITCHES.

receive the hook of an ox-chain. This scraper may be used for any straight-forward work by attaching the power to the middle of the chain. By moving the hook a few links to the right or left, it will act somewhat after the manner of the mould-board of a plow, and will, if skillfully handled, shoot the filling rapidly into the ditch.

If the work is done by hand, mix the surface soil and turf with the subsoil filling for the whole depth. If with a scraper, put the surface soil at the bottom of the loose filling, and the subsoil at the top, as this will be an imitation, for the limited area of the drains, of the process of "trenching," which is used in garden cultivation.

When the ditches are filled, they will be higher than the adjoining land, and it will be well to make them still more so by digging or plowing out a small trench at each side of the drain, throwing the earth against the mound, which will prevent surface water, (during heavy rains,) from running into the loose filling before it is sufficiently settled. A cross section of a filled drain provided with these ditches is shown in Figure 40.

In order that the silt-basins may be examined, and their accumulations of earth removed, during the early action of the drains, those parts of the ditches which are above them may be left open, care being taken, by cutting surface ditches around them, to prevent the entrance of water from above. During this time the covers of the basins should be kept on, and should be covered with inverted sods to keep loose dirt from getting into them.

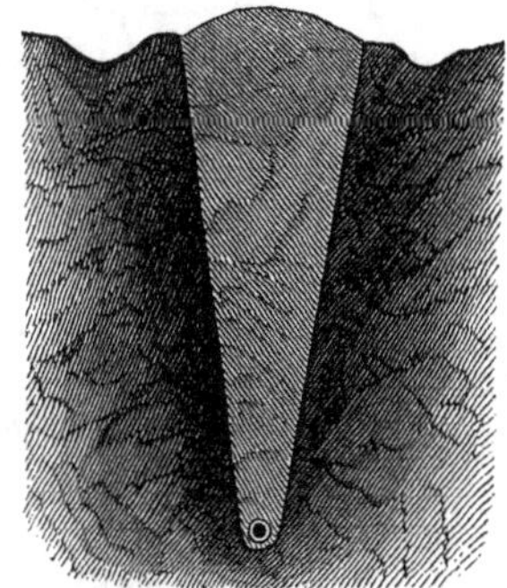

Fig. 40.—CROSS-SECTION OF DITCH (FILLED), WITH FURROW AT EACH SIDE.

Collecting the Water of Springs.—The lateral which connects with the main drain, *A*, (Fig. 21,) at the point *m*, and which is to take the water of the spring at the head of the brook, should not be opened until the main has been completed and filled into the silt-basin,—the brook having, meantime, been carried over the other ditches in wooden troughs. This lateral may now be made in the following way: Dig down to the tile of the main, and carry the lateral ditch back, a distance of ten feet. In the bottom of this, place a wooden trough, at least six feet long, laid at such depth that its channel shall be on the exact grade required for laying the tiles, and lay long straw, (held down by weights,) lengthwise within it. Make an opening in the tile of the main and connect the trough with it. The straw will prevent any coarse particles of earth from being carried into the tile, and the flow of the water will be sufficient to carry on to the silt-basin any finer matters. Now open the ditch to and beyond the spring, digging at least a foot below the grade in its immediate vicinity, and filling to the exact grade with small stones, broken bricks, or other suitable material. Lay the tiles from the upper end of the ditch across the stone work, and down to the wooden trough. Now spread a sufficient layer of wood shavings over the stone work to keep the earth from entering it, cover the tiles and fill in the ditch, as before directed, and then remove the straw from the wooden trough and lay tiles in its place. In this way, the water of even a strong spring may be carried into a finished drain without danger. In laying the tile which crosses the stone work, it is well to use full 2½-inch tiles in the place of collars, leaving the joints of these, and of the 1¼-inch tiles, (which should join near the middle of the collar tile,) about a quarter of an inch open, to give free entrance to the water.

The stone and tile drain, *H*, *I*, is simply dug out to the surface of the rock, if this is not more than two feet below the grade of the upper ends of the laterals with which it connects, and then filled up with loose stones to the line of grade. If the stones are small, so as to form a good bottom for the tiles, they may be laid directly upon it; if not, a bottom for them may be made of narrow strips of cheap boards. Before filling, the tiles and stone work should be covered with shavings, and the filling above these should consist of a strong clay, which will remain in place after the shavings rot away.

Amending the Map.—When the tiles are laid, and before they are covered, all deviations of the lines, as in passing around large stones and other obstructions, which may have prevented the exact execution of the original plan, and the location and kind of each underground silt-basin should also be carefully noted, so that they may be transferred to the map, for future reference, in the event of repairs becoming necessary. In a short time after the work is finished, the surface of the field will show no trace of the lines of drain, and it should be possible, in case of need, to find any point of the drains with precision, so that no labor will be lost in digging for it. It is much cheaper to measure over the surface than to dig four feet trenches through the ground.

CHAPTER V.

HOW TO TAKE CARE OF DRAINS AND DRAINED LAND.

So far as tile drains are concerned, if they are once well laid, and if the silt-basins have been emptied of silt until the water has ceased to deposit it, they need no care nor attention, beyond an occasional cleaning of the outlet brook. Now and then, from the proximity of willows, or thrifty, young, water-loving trees, a drain will be obstructed by roots; or, during the first few years after the work is finished, some weak point,—a badly laid tile, a loosely fitted connection between the lateral and a main, or an accumulation of silt coming from an undetected and persistent vein of quicksand,—will be developed, and repairs will have to be made. Except for the slight danger from roots, which must always be guarded against to the extent of allowing no young trees of the dangerous class to grow near a drain through which a *constant* stream of water flows, it may be fairly assumed that drains which have been kept in order for four or five years have passed the danger of interruption from any cause, and they may be considered entirely safe.

A drain will often, for some months after it is laid, run muddy water after rains. Sometimes the early deposit of silt will nearly fill the tile, and it will take the water of several storms to wash it out. If the tiles have been laid in packed clay, they cannot long receive silt from without, and that which makes the flow turbid, may be assumed to come from the original deposit in the conduit. Examinations of newly laid drains have developed many instances where tiles were at first half filled with silt, and three months later were entirely clean. The muddiness of the flow indicates what the doctors call "an effort of nature to relieve herself," and nature may be trusted to succeed, at least, until she abandons the effort. If we are sure that a drain has been well laid, we need feel no anxiety because it fails to take the water from the ground so completely as it should do, until it settles into a flow of clear water after the heaviest storms.

In the case of an actual stoppage, which will generally be indicated by the "bursting out" of the drain, i. e., the wetting of the land as though there were a spring under it, or as though its water had no underground outlet, (which is the fact,) it will be necessary to lay open the drain until the obstruction is found.

In this work, the real value of the map will be shown, by the facility which it offers for finding any point of any line of drains, and the exact locality of the junctions with the mains, and of the silt-basins. In laying out the plan on the ground, and in making his map, the surveyor will have had recourse to two or more fixed points; one of them, in our example, (fig. 21,) would probably be the center of the main silt-basin, and one, a drilled hole or other mark on the rock at the north side of the field. By staking out on the ground the straight line connecting these two points, and drawing a corresponding line on the map, we shall have a *base-line*, from which it will be easy, by perpendicular offsets, to determine on the ground any point upon the map. By laying a small square on the map, with one of its edges coinciding with the base-line, and moving it on this line until the other edge meets the desired point, we fix, at the angle of the square, the point on the base-line from which we are to measure the length of the offset. The next step is to find, (by the scale,) the distance of this point from the nearest end of the base-line, and from the point sought. Then measure off, in the field, the corresponding distance on the base-line, and, from the point thus found, measure on a line perpendicular to the base line, the length of the offset; the point thus indicated will be the locality sought. In the same manner, find another point on the same drain, to give the range on which to stake it out. From this line, the drains which run parallel to it, can easily be found, or it may be used as a base-line, from which to find, by measuring offsets, other points near it.

The object of this staking is, to find, in an inexpensive and easy way, the precise position of the drains, for which it would be otherwise necessary to grope in the dark, verifying our guesses by digging four-foot trenches, at random.

If there is a silt-basin, or a junction a short distance below the point where the water shows itself, this will be the best place to dig. If it is a silt-basin, we shall probably find that this has filled up with dirt, and has stopped the flow. In this case it should be cleaned out, and a point of the drain ten feet below it examined. If this is found to be clear, a long slender stick may be pushed up as far as the basin and worked back and forth until the passage is cleared. Then replace the tile below, and try with the stick to clean the tiles above the basin, so as to tap the water above the obstruction. If this cannot be done, or if the drain ten feet below is clogged, it will be necessary to uncover the tiles in both directions until an opening is found, and to take up and relay the whole. If the wetting of the ground is sufficient to indicate that there is much water in the drain, only five or six tiles should be taken up at a time, cleaned and relaid,—commencing at the lower end,—in order that, when the water commences to flow, it may not disturb the bottom of the ditch for the whole distance.

If the point opened is at a junction with the main, examine both the main and the lateral, to see which is stopped, and proceed with one or the other, as directed above. In doing this work, care should be taken to send as little muddy water as possible into the drain below, and to allow the least possible disturbance of the bottom.

If silt-basins have been placed at those points at which the fall diminishes, the obstruction will usually be found to occur at the outlets of these, from a piling up of the silt in front of them, and to extend only a short distance below and above. It is not necessary to take up the tiles until they are found to be entirely clean, for, if they are only one-half or one-third full, they will probably be washed clean by the rush of water, when that which is accumulated above is tapped. The work should be done in settled fair weather, and the ditches should remain open until the effect of the flow has been observed. If the tiles are made thoroughly clean by the time that the accumulated water has run off, say in 24 hours, they may be covered up; if not, it may be necessary to remove them again, and clean them by hand. When the work is undertaken it should be thoroughly done, so that the expense of a new opening need not be again incurred.

It is worse than useless to substitute larger sizes of tiles for those which are taken up. The obstruction, if by silt, is the result of a too sluggish flow, and to enlarge the area of the conduit would only increase the difficulty. If the tiles are too small to carry the full flow which follows a heavy rain, they will be very unlikely to become choked, for the water will then have sufficient force to wash them clean, while if they are much larger than necessary, a deposit of silt to one half of their height will make a broad, flat bed for the stream, which will run with much less force, and will be more likely to increase the deposit.

If the drains are obstructed by the roots of willows, or other trees, the proprietor must decide whether he will sacrifice the trees or the drains; both he cannot keep, unless he chooses to go to the expense of laying in cement all of the drains which carry constant streams, for a distance of at least 50 feet from the dangerous trees. The trouble from trees is occasionally very great, but its occurrence is too rare for general consideration, and must be met in each case with such remedies as circumstances suggest as the best.

The gratings over the outlets of silt-basins which open at the surface of the ground, are sometimes, during the first year of the drainage, obstructed by a fungoid growth which collects on the cross bars. This should be occasionally rubbed off. Its character is not very well understood, and it is rarely observed in old drains. The decomposition of the grass bands which are used to cover the joints of the larger tiles may encourage its formation.

If the surface soil have a good proportion of sand, gravel, or organic matter, so as to give it the consistency which is known as "loamy," it will bear any treatment which it may chance to receive in cultivation, or as pasture land; but if it be a decided clay soil, no amount of draining will enable us to work it, or to turn cattle upon it when it is wet with recent rains. It will much sooner become dry, because of the drainage, and may much sooner be trodden upon without injury; but wet clay cannot be worked or walked over without being more or less *puddled*, and, thereby, injured for a long time.

No matter how thoroughly heavy clay pasture lands may be under-drained, the cattle should be removed from them when it rains, and kept off until they are comparatively dry. Neglect of this precaution has probably led to more disappointment as to the effects of drainage than any other circumstances connected with it. The injury from this cause does not extend to a great depth, and in the Northern States it would always be overcome by the frosts of a single winter; as has been before stated, it is confined to stiff clay soils, but as these are the soils which most need draining, the warning given is important.

CHAPTER VI.

WHAT DRAINING COSTS.

Draining is expensive work. This fact must be accepted as a very stubborn one, by every man who proposes to undertake the improvement. There is no royal road to tile-laying, and the beginner should count the cost at the outset. A good many acres of virgin land at the West might be bought for what must be paid to get an efficient system of drains laid under a single acre at home. Any man who stops at this point of the argument will probably move West,—or do nothing.

Yet, it is susceptible of demonstration that, even at the West, in those localities where Indian Corn is worth as much as fifty cents per bushel at the farm, it will pay to drain, in the best manner, all such land as is described in the first chapter of this book as in need of draining, Arguments to prove this need not be based at all on cheapness of the work; only on its effects and its permanence.

In fact, so far as draining with tiles is concerned, cheapness is a delusion and a snare, for the reason that it implies something less than the best work, a compromise between excellence and inferiority. The moment that we come down from the best standard, we introduce a new element into the calculation. The sort of tile draining which it is the purpose of this work to advocate is a system so complete in every particular, that it may be considered as an absolutely permanent improvement. During the first years of the working of the drains, they will require more or less attention, and some expense for repairs; but, in well constructed work, these will be very slight, and will soon cease altogether. In proportion as we resort to cheap devices, which imply a neglect of important parts of the work, and a want of thoroughness in the whole, the expense for repairs will increase, and the duration of the usefulness of the drains will diminish.

Drains which are permanently well made, and which will, practically, last for all time, may be regarded as a good investment, the increased crop of each year, paying a good interest on the money that they cost, and the money being still represented by the undiminished value of the improvement. In such a case the draining of the land may be said to cost, not $50 per acre,—but the interest on $50 each year. The original amount is well invested, and brings its yearly dividend as surely as though it were represented by a five-twenty bond.

With badly constructed drains, on the other hand, the case is quite different. In buying land which is subject to no loss in quantity or quality, the farmer considers, not so much the actual cost, as the relation between the yearly interest on the cost, and the yearly profit on the crop,—knowing that, a hundred years hence, the land will still be worth his money.

But if the land were bounded on one side by a river which yearly encroached some feet on its bank, leaving the field a little smaller after each freshet; or if, every spring, some rods square of its surface were sure to be covered three feet deep with stones and sand, so that the actual value of the property became every year less, the purchaser would compare the yearly value of the crops, not only with the interest on the price, but, in addition to this, with so much of the prime value as yearly disappears with the destruction of the land.

It is exactly so with the question of the cost of drainage. If the work is insecurely done, and is liable, in five years or in fifty, to become worthless; the increase of the crops resulting from it, must not only cover the yearly interest on the cost, but the yearly depreciation as well. Therefore what may seem at the time of doing the work to be cheapness, is really the greatest extravagance. It is like buiding a brick wall with clay for mortar. The bricks and the workmanship cost full price, and the small saving on the mortar will topple the wall over in a few years, while, if well cemented, it would have lasted for centuries. The cutting and filling of the ditches, and the purchase and transportation of the tiles, will cost the same in every case, and these constitute the chief cost; if the proper care in grading, tile-laying and covering, and in making outlets be stingily withheld,—saving, perhaps, one-tenth of the expense,—what might have been a permanent improvement to the land, may disappear, and the whole outlay be lost in ten years. A saving of ten per cent. in the cost will have lost us the other ninety in a short time.

But, while cheapness is to be shunned, economy is to be sought in every item of the work of draining, and should be studied, by proprietor and engineer, from the first examination of the land, to the throwing of the last shovelful of earth on to the filling of the ditch. There are few operations connected with the cultivation of the soil in which so much may be imperceptibly lost through neglect, and carelessness about little details, as in tile-draining. In the original levelling of the ground, the adjustment of the lines, the establishing of the most judicious depth and inclination at each point of the drains, the disposition of surface streams during the prosecution of the work, and in the width of the excavation, the line which divides economy and wastefulness is extremely narrow, and the most constant vigilance, together with the best judgment and foresight, are needed to avoid unnecessary cost. In the laying and covering of the tile, on the other hand, it is best to disregard a little slowness and unnecessary care on the part of the workmen, for the sake of the most perfect security of the work.

Details of Cost.—The items of the work of drainage may be classified as follows:

1. Engineering and Superintendence.
2. Digging the ditches.
3. Grading the bottoms.
4. Tile and tile-laying.
5. Covering the tile and filling the ditches.
6. Outlets and silt-basins.

1. *Engineering and Superintendence.*—It is not easy to say what would be the proper charge for this item of the work. In England, the Commissioners under the Drainage Acts of Parliament, and the Boards of Public Works, fix the charge for engineering at $1.25 per acre. That is in a country when the extent of lands undergoing the process of draining is very great, enabling one person to superintend large tracts in the same neighborhood at the same time, and with little or no outlay for travelling expenses. In this country, where the improvement is, thus far, confined to small areas, widely separated; and where there are comparatively few engineers who make a specialty of the work, the charge for services is necessarily much higher, and the amount expended in travelling much greater. In most cases, the proprietor of the land must qualify himself to superintend his own operations, (with the aid of a country surveyor, or a railroad engineer in the necessary instrumental work.) As draining becomes more general, the demand for professional assistance will, without doubt, cause local engineers to turn their attention to the subject, and their services may be more cheaply obtained. At present, it would probably not be prudent to

estimate the cost of engineering and superintendence, including the time and skill of the proprietor, at less than $5 per acre, even where from 20 to 50 acres are to be drained at once.

2. *Digging the Ditches.*—The labor required for the various operations constitutes the principal item of cost in draining, and the price of labor is now so different in different localities, and so unsettled in all, that it is difficult to determine a rate which would be generally fair. It will be assumed that the average wages of day laborers of the class employed in digging ditches, is $1.50 per day, and the calculation will have to be changed for different districts, in proportion to the deviation of the actual rate of wages from this amount. There is a considerable advantage in having the work done at some season, (as after the summer harvest, or late in the fall,) when wages are comparatively low.

The cutting of the ditches should always be let by the rod. When working at day's work, the men will invariably open them wider than is necessary, for the sake of the greater convenience of working, and the extra width causes a corresponding waste of labor.

A 4-foot ditch, in most soils, need be only 20 inches wide at the surface, and 4 inches at the bottom. This gives a mean width of 12 inches, and requires the removal of nearly 2½ cubic yards of earth for each rod of ditch; but an increase to a mean width of 16 inches, (which day workmen will usually reach, while piece workmen almost never will,) requires the removal of 3¼ cubic yards to the rod. As the increased width is usually below the middle of the drain, the extra earth will all have to be raised from 2 to 4 feet, and the extra ¾ yards will cost as much as a full yard taken evenly from the whole side, from top to bottom.

In clay soils, free from stones or "hard pan," but so stiff as to require considerable picking, ordinary workmen, after a little practice, will be able to dig 3½ rods of ditch per day, to an average depth of 3.80,—leaving from 2 to 3 inches of the bottom of 4-foot ditches to be finished by the graders. This makes the cost of digging about 43 cents per rod. In loamy soil the cost will be a little less than this, and in very hard ground, a little more. In sandy and peaty soils, the cost will not be more than 30 cents. Probably 43 cents would be a fair average for soils requiring drainage, throughout the country.

This is about 17 cents for each yard of earth removed.

In soft ground, the caving in of the banks will require a much greater mean width than 12 inches to be thrown out, and, if the accident could not have been prevented by ordinary care on the part of the workman, (using the bracing boards shown in Fig. 28,) he should receive extra pay for the extra work. In passing around large stones it may also be necessary to increase the width.

The following table will facilitate the calculations for such extra work:

CUBIC YARDS OF EXCAVATION IN DITCHES OF VARIOUS WIDTH.

Length of Ditch.	12 *Inches Wide.*		18 *Inches Wide.*		24 *Inches Wide.*		30 *Inches Wide.*		36 *Inches Wide.*	
	Yds.	Feet.	Yds.	Feet.	Yds.	Feet.	Yds.	Feet.	Yds.	Feet.
1 Yard..........		12		18		24	1	3	1	9
1 Rod...........	2	12	3	18	4	24	6	3	7	9

Men will, in most soils, work best in couples,—one shovelling out the earth, and working forward, and the other, (moving backward,) loosening the earth with a spade or foot-pick, (Fig. 41.) In stony land, the men should be required to keep their work well closed up,—excavating to the full depth as they go. Then, if they strike a stone too large to be taken out within the terms of their contract, they can skip a sufficient distance to pass it, and the digging of the omitted part may be done by a faithful day workman. This will usually be cheaper and more satisfactory than to pay the contractors for extra work.

Concerning the amount of work that one man can do in a day, in different soils, digging ditches 4 feet deep, French says: "In the writer's own field, " where the pick was used to loosen the lower " two feet of earth, the labor of opening and "*filling* drains 4 feet deep, and of the mean " width of 14 inches, all by hand labor, has " been, in a mile of drains, being our first ex- " periments, about one day's labor to 3 rods " in length. The excavated earth of such a " drain measures not quite 3 cubic yards, " (exactly, 2.85.)" In a subsequent work, in a sandy soil, two men opened, *laid*, and *refilled* 14 rods in one day;—the mean width being 12 inches.*

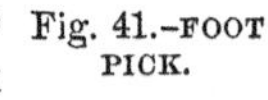
Fig. 41.-FOOT PICK.

" In the same season, the same men opened, "*laid*, and *filled* 70 rods of 4-foot drain of " the same mean width of 12 inches, in the " worst kind of clay soil, where the pick " was constantly used. It cost 35 days' labor to complete " the job, being 50 cents per rod for the labor alone." Or, under the foregoing calculation of $1.50 per day, 75 cents per rod. These estimates, in common with nearly all that are published, are for the entire work of digging, grading, tile-laying, and refilling. Deducting the time required for the other work, the result will be about as above estimated; for the rough excavation, 3½ rods to the day's work, costing, at $1.50 per day, 43 cents to the rod.

Grading is the removal of 2 or 3 inches in depth, and about 4 inches in width, of the soil at the bottom of the ditch. It is chiefly done with the finishing scoop, which, (being made of two thin plates, one of iron and one of steel, welded together, the iron wearing away and leaving

* Surely such soil ought not to require thorough draining; where men can go so easily, water ought to find its way alone.

the sharp steel edge always prominent,) will work in a very hard clay without the aid of the pick. Three men,—the one in the ditch being a skillful workman, and the others helping him when not sighting the rods,—will grade about 100 rods per day, making the cost about 6 cents per rod. Until they acquire the skill to work thus rapidly, they should not be urged beyond what they can readily do in the best manner, as this operation, (which is the preparing of the foundation for the tiles,) is probably the most important of the whole work of draining.

Tiles and Tile-Laying.—After allowing for breakage, it will take about 16 tiles and 16 collars to lay a rod in length of drain. The cost of these will, of course, be very much affected by the considerations of the nearness of the tile-kiln and the cost of transportation. They should, in no ordinary case, cost, delivered on the ground, more than $8 per thousand for 1¼-inch tiles, and $4 per thousand for the collars, making a total of $12 for both, equal to about 19 cents per rod. The laying of the tiles, may be set down at 2 cents per rod,—based on a skilled man laying 100 rods daily, and receiving $2 per day.

Covering and filling will probably cost 10 cents per rod, (if the scraper, Fig. 39, can be successfully used for the rough filling, the cost will be reduced considerably below this.)

The four items of the cost of making one rod of lateral drain are as follows:

Digging the ditches - - - - - - - - -	.43
Grading - - - - - - - - - - - - - -	.06
Tiles and laying - - - - - - - - - -	.21
Covering and filling - - - - - - - - -	.10
	.80 cts.

If the drains are placed at intervals of 40 feet, there are required 64 rods to the acre,—this at 80 cents per rod will make the cost per acre,—for the above items,—$51.20.

How much should be allowed for main drains, outlets, and silt-basins, it is impossible to say, as, on irregular ground, no two fields will require the same amount of this sort of work. On very even land, where the whole surface, for hundreds of acres, slopes gradually in one or two directions, the outlay for mains need not be more than two per cent. of the cost of the laterals. This would allow laterals of a uniform length of 800 feet to discharge into the main line, at intervals of 40 feet, if we do not consider the trifling extra cost of the larger tiles. On less regular ground, the cost of mains will often be considerably more than two per cent. of the cost of the laterals; but in some instances the increase of main lines will be fully compensated for by the reduction in the length of the laterals, which, owing to rocks, hills too steep to need drains at regular intervals, and porous, (gravelly,) streaks in the land, cannot be profitably made to occupy the whole area so thoroughly.*

Probably 7½ per cent. of the cost of the laterals for mains, outlets, and silt-basins will be a fair average allowance.

This will bring the total cost of the work to about $60 per acre, made up as follows:

Cost of the finished drains per acre - -	$51.20
7½ per cent. added for mains, etc. - - -	3.83
Engineering and Superintendence - - -	5.00

Of course this is an arbitrary calculation, an estimate without a single ascertained fact to go upon,—but it is as

* The land shown in Fig. 21, is especially irregular, and, for the purpose of illustrating the principles upon which the work should be done, an effort has been made to make the work as complete as possible in all particulars. In actual work on a field similar to that, it would not probably be good economy to make all the drains laid in the plan, but as deviations from the plan would depend on conditions which cannot well be shown on such a small scale, they are disregarded, and the system of drains is made as it would be if it were all plain sailing.

close as it can be made to what would probably be the cost of the best work, on average ground, at the present high prices of labor and material. Five years ago the same work could have been done for from $40 to $45 per acre, and it will be again cheaper when wages fall, and when a greater demand for draining tiles shall have caused more competition in their manufacture. With a large general demand, such as has existed in England for the last 20 years, they would now be sold for one-half of their present price here, and the manufacture would be more profitable.

There are many light lands on retentive subsoils, which could be drained, at present prices, for $50 or less per acre, and there are others, which are very hard to dig, on which thorough-draining could not now be done for $60.

The cost and the promise of the operation in each instance, must guide the land owner in deciding whether or not to undertake the improvement.

In doubtful cases, there is one compromise which may be safely made,—that is, to omit each alternate drain, and defer its construction until labor is cheaper.

This is doing half the work,—a very different thing from half-doing the work. In such cases, the lines should be laid out as though they were to be all done at once, and, finally, when the omitted drains are made, it should be in pursuance of the original plan. Probably the drains which are laid will produce more than one-half of the benefit that would result if they were all laid, but they will rarely be satisfactory, except as a temporary expedient, and the saving will be less than would at first seem likely, for when the second drains are laid; the cultivation of the land must be again interrupted; the draining force must be again brought together; the levels of the new lines must be taken, and connected with those of the old ones; and great care must be taken, selecting the dryest weather for the work,—to admit very little, if any, muddy water into the old mains.

This practice of draining by installments is not recommended; it is only suggested as an allowable expedient, when the cost of the complete work could not be borne without inconvenience.

If any staid and economical farmer is disposed to be alarmed at the cost of draining, he is respectfully reminded of the miles of expensive stone walls and other fences, in New England and many other parts of the country, which often are a real detriment to the farms, occupying, with their accompanying bramble bushes and head lands, acres of valuable land, and causing great waste of time in turning at the ends of short furrows in plowing;—while they produce no benefit at all adequate to their cost and annoyance.

It should also be considered that, just as the cost of fences is scarcely felt by the farmer, being made when his teams and hands could not be profitably employed in ordinary farming operations, so the cost of draining will be reduced in proportion to the amount of the work which he can "do within himself,"—without hiring men expressly for it. The estimate herein given is based on the supposition that men are hired for the work, at wages equal to $1.50 per day,—while draining would often furnish a great advantage to the farmer in giving employment to farm hands who are paid and subsisted by the year.

CHAPTER VII.

"WILL IT PAY?"

Starting with the basis of $60, as the cost of draining an acre of ordinary farm land;—what is the prospect that the work will prove remunerative?

In all of the older States, farmers are glad to lend their surplus funds, on bond and mortgage on their neighbors' farms, with interest at the rate of 7, and often 6 per cent.

In view of the fact that a little attention must be given each year to the outlets, and, to the silt-basins, as well, for the first few years, it will be just to charge for the use of the capital 8⅓ per cent.

This will make a yearly charge on the land, for the benefits resulting from such a system of draining as has been described, OF FIVE DOLLARS PER ACRE.

Will it Pay?—Will the benefits accruing, year after year,—in wet seasons and in dry,—with root crops and with grain,—with hay and with fruit,—in rotations of crops and in pasture,—be worth $5 an acre?

On this question depends the value of tile-draining as a *practical* improvement, for if there is a self-evident proposition in agriculture, it is that what is not profitable, one year with another, is *not* practical.

To counterbalance the charge of $5, as the yearly cost

of the draining, each acre must produce, in addition to what it would have yielded without the improvement:

10	bushels	of	Corn	at	.50	per bushel.
3	"	"	Wheat	"	$1.66	" "
5	"	"	Rye	"	1.00	" "
12½	"	"	Oats	"	.40	" "
10	"	"	Potatoes	"	.50	" "
6⅔	"	"	Barley	"	.75	" "
1,000	pounds	"	Hay	"	10.00	" ton.
50	"	"	Cotton	"	.10	" pound.
20	"	"	Tobacco	"	.25	" "

Surely this is not a large increase,—not in a single case,—and the prices are generally less than may be expected for years to come.

The United States Census Report places the average crop of Indian Corn, in Indiana and Illinois, at 33 bushels per acre. In New York it was but 27 bushels, and in Pennsylvania but 20 bushels. It would certainly be accounted extremely liberal to fix the average yield of such soils as need draining, at 30 bushels per acre. It is extremely unlikely that they would yield this, in the average of seasons, with the constantly recurring injury from backward springs, summer droughts, and early autumn frosts.

Heavy, retentive soils, which are cold and late in the spring, subject to hard baking in midsummer, and to become cold and wet in the early fall, are the very ones which are best suited, when drained, to the growth of Indian Corn. They are "strong" and fertile,—and should be able to absorb, and to prepare for the use of plants, the manure which is applied to them, and the fertilizing matters which are brought to them by each storm;—but they cannot properly exercise the functions of fertile soils, for the reason that they are strangled with water, chilled by evaporation, or baked to almost brick-like hardness, during nearly the whole period of the growth and ripening of the crop.

The manure which has been added to them, as well as their own chemical constituents, are prevented from undergoing those changes which are necessary to prepare them for the uses of vegetation. The water of rains, finding the spaces in the soil already occupied by the water of previous rains, cannot enter to deposit the gases which it contains,—or, if the soil has been dried by evaporation under the influence of sun and wind, the surface is almost hermetically sealed, and the water is only slowly soaked up, much of it running off over the surface, or lying to be removed by the slow and chilling process of evaporation. In wet times and in dry, the air, with its heat, its oxygen, and its carbonic acid, (its universal solvent,) is forbidden to enter and do its beneficent work. The benefit resulting from cultivating the surface of the ground is counteracted by the first unfavorable change of the weather; a single heavy rain, by saturating the soil, returning it to nearly its original condition of clammy compactness. In favorable seasons, these difficulties are lessened, but man has no control over the seasons, and to-morrow may be as foul as to-day has been fair. A crop of corn on undrained, retentive ground, is subject to injury from disastrous changes of the weather, from planting until harvest. Even supposing that, in the most favorable seasons, it would yield as largely as though the ground were drained, it would lose enough in unfavorable seasons to reduce the average more than ten (10) bushels per acre.

The average crop, on such land, has been assumed to be 30 bushels per acre; it would be an estimate as moderate as this one is generous, to say that, with the same cultivation and the same manure, the average crop, after draining, would be 50 bushels, or an increase equal to twice as much as is needed to pay the draining charge. If the method of cultivation is improved, by deep plowing, ample manuring, and thorough working,—all of which may be more profitably applied to drained than to undrained land,—the *average* crop,—of a series of years,—will not be less than 60 bushels.

The cost of extra harvesting will be more than repaid by the value of the extra fodder, and the increased cultivation and manuring are lasting benefits, which can be charged, only in small part, to the current crop. Therefore, if it will pay to plow, plant, hoe and harvest for 30 bushels of corn, it will surely pay much better to double the crop at a yearly extra cost of $5, and, practically, it amounts to this;—the extra crop is nearly all clear gain.

The quantity of Wheat required to repay the annual charge for drainage is so small, that no argument is needed to show that any process which will simply prevent "throwing out" in winter, and the failure of the plant in the wetter parts of the field, will increase the product more than that amount,—to say nothing of the general importance to this crop of having the land in the most perfect condition, (in winter as well as in summer.)

It is stated that, since the general introduction of drainage in England, (within the past 25 years,) the wheat crop of that country has been more than doubled. Of course, it does not necessarily follow that the amount *per acre* has been doubled, large areas which were originally unfit for the growth of this crop, having been, by draining, excellently fitted for its cultivation;—but there can be no doubt that its yield has been greatly increased on all drained lands, nor that large areas, which, before being drained, were able to produce fair crops only in the best seasons, are now made very nearly independent of the weather.

It is not susceptible of demonstration, but it is undoubtedly true, that those clay or other heavy soils, which are devoted to the growth of wheat in this country, would, if they were thoroughly under-drained, produce, on the average of years, at least double their present crop.

Mr. John Johnston, a venerable Scotch farmer, who has

long been a successful cultivator in the Wheat region of Western New York,—and who was almost the pioneer of tile-draining in America,—has laid over 50 miles of drains within the last 30 years. His practice is described in Klippart's Land Drainage, from which work we quote the following:

"Mr. Johnston says he never saw 100 acres in any one "farm, but a portion of it would pay for draining. Mr. "Johnston is no rich man who has carried a favorite hobby "without regard to cost or profit. He is a hardworking "Scotch farmer, who commenced a poor man, borrowed "money to drain his land, has gradually extended his "operations, and is now reaping the benefits, in having "crops of 40 bushels of wheat to the acre. He is a gray-"haired Nestor, who, after accumulating the experience "of a long life, is now, at 68 years of age, written to by "strangers in every State of the Union for information, "not only in drainage matters, but all cognate branches "of farming. He sits in his homestead, a veritable Hum-"boldt in his way, dispensing information cheerfully "through our agricultural papers and to private corres-"pondents, of whom he has recorded 164 who applied to "him last year. His opinions are, therefore, worth more "than those of a host of theoretical men, who write with-"out practice." * * * * * * * * * * * *

"Although his farm is mainly devoted to wheat, yet a "considerable area of meadow and some pasture has been "retained. He now owns about 300 acres of land. The "yield of wheat has been 40 bushels this year, and in for-"mer seasons, when his neighbors were reaping 8, 10, or "15 bushels, he has had 30 and 40." * * * * *

"Mr. Johnston says tile-draining pays for itself in two "seasons, sometimes in one. Thus, in 1847, he bought a "piece of 10 acres to get an outlet for his drains. It was "a perfect quagmire, covered with coarse aquatic grasses, "and so unfruitful that it would not give back the seed

" sown upon it. In 1848 a crop of corn was taken from it, " which was measured and found to be *eighty bushels* per " acre, and as, because of the Irish famine, corn was worth " $1 per bushel that year, this crop paid not only all the ex- " pense of drainage, but the first cost of the land as well.

" Another piece of 20 acres, adjoining the farm of the " late John Delafield, was wet, and would never bring " more than 10 bushels of corn per acre. This was drained " at a great cost, nearly $30 per acre. The first crop after " this was 83 bushels and some odd pounds per acre. It " was weighed and measured by Mr. Delafield, and the " County Society awarded a premium to Mr. Johnston. " Eight acres and some rods of this land, at one side, aver- " aged 94 bushels, or the trifling increase of 84 bushels " per acre over what it would bear before those insignifi- " cant clay tiles were buried in the ground. But this in- " crease of crop is not the only profit of drainage; for Mr. " Johnston says that, on drained land, one half the usual " quantity of manure suffices to give maximum crops. It " is not difficult to find a reason for this. When the soil " is sodden with water, air can not enter to any extent, " and hence oxygen can not eat off the surfaces of soil- " particles and prepare food for plants; thus the plant " must in great measure depend on the manure for suste- " nance, and, of course, the more this is the case, the more " manure must be applied to get good crops. This is one " reason, but there are others which we might adduce if " one good one were not sufficient.

" Mr. Johnston says he never made money until he " drained, and so convinced is he of the benefits accruing " from the practice, that he would not hesitate,—as he did " not when the result was much more uncertain than at " present,—to borrow money to drain. Drains well laid, " endure, but unless a farmer intends doing the job well, " he had best leave it alone and grow poor, and move out " West, and all that sort of thing. Occupiers of appar-

" ently dry land are not safe in concluding that they need " not go to the expense of draining, for if they will but " dig a three-foot ditch in even the driest soil, water will " be found in the bottom at the end of eight hours, and " if it does come, then draining will pay for itself " speedily."

Some years ago, the Rural New Yorker published a letter from one of its correspondents from which the following is extracted: —

" I recollect calling upon a gentleman in the harvest field, when something like the following conversation occurred:

'Your wheat, sir, looks very fine; how many acres have you in this field?'

'In the neighborhood of eight, I judge.'

'Did you sow upon fallow?'

'No sir. We turned over green sward—sowed immediately upon the sod, and dragged it thoroughly—and you see the yield will probably be 25 bushels to the acre, where it is not too wet.'

'Yes sir, it is mostly very fine. I observed a thin strip through it, but did not notice that it was wet.'

'Well, it is not *very* wet. Sometimes after a rain, the water runs across it, and in spring and fall it is just wet enough to heave the wheat and kill it.'

I inquired whether a couple of good drains across the lot would not render it dry.

'Perhaps so—but there is not over an acre that is killed out.'

'Have you made an estimate of the loss you annually sustain from this wet place?'

'No, I had not thought much about it.'

'Would $30 be too high?'

'O yes, double.'

'Well, let's see; it cost you $3 to turn over the sward? Two bushels of seed, $2; harrowing in, 75 cents; interest, taxes, and fences, $5.25; 25 bushels of wheat lost, $25.'

'Deduct for harvesting———'

'No; the straw would pay for that.'

'Very well, all footed $36.'

'What will the wheat and straw on this acre be worth this year?'

'Nothing, as I shall not cut the ground over.'

'Then it appears that you have lost, in what you have actually expended, and the wheat you would have harvested, had the ground been dry, $36, a pretty large sum for one acre.'

'Yes I see,' said the farmer."

While Rye may be grown, with tolerable advantage, on lands which are less perfectly drained than is necessary for Wheat, there can be no doubt that an increase of more than the six and two-thirds bushels needed to make up the drainage charge will be the result of the improvement.

While Oats will thrive in soils which are too wet for many other crops, the ability to plant early, which is secured by an early removal from the soil of its surplus water, will ensure, one year with another, more than twelve and a half bushels of increased product.

In the case of Potatoes, also, the early planting will be a great advantage; and, while the cause of the potato-rot is not yet clearly discovered, it is generally conceded that, even if it does not result directly from too great wetness of the soil, its development is favored by this condition, either from a direct action on the tubers, or from the effect in the air immediately about the plants, of the exhalations of a humid soil.

An increase of from five to ten per cent. on a very ordinary crop of potatoes, will cover the drainage charge, and, with facilities for marketing, the higher price of the earlier yield is of much greater consequence.

Barley will not thrive in wet soil, and there is no question that drainage would give it much more than the increased yield prescribed above.

As to hay, there are many wet, rich soils which produce very large crops of grass, and it is possible that drainage might not always cause them to yield a thousand pounds more of hay to the acre, but the *quality* of the hay from the drained soil, would, of itself, more than compensate for the drainage charge. The great benefit of the improvement, with reference to this crop, however, lies in the fact that, although wet, grass lands,—and by "wet" is meant the condition of undrained, retentive clays, and heavy loams, or other soils requiring drainage,—in a very few years "run out," or become occupied by semi-aquatic and other objectionable plants, to the exclusion of the proper grasses; the same lands, thoroughly drained, may be kept in full yield of the finest hay plants, as long as the ground is properly managed. It must, of course, be manured, from time to time, and care should be taken to prevent the puddling of its surface, by men or animals, while it is too wet from recent rain. With proper attention to these points, it need not be broken up in a lifetime, and it may be relied on to produce uniformly good crops, always equal to the best obtained before drainage.

So far as Cotton and Tobacco are concerned, there are not many instances recorded of the systematic drainage of lands appropriated to their cultivation, but there is every reason to suppose that they will both be benefitted by any operation which will have the effect of placing the soil in a better condition for the uses of all cultivated plants. The average crop of tobacco is about 700 lbs., and that of cotton probably 250 lbs. An addition of one-fifth to the cotton crop, and of only one thirty-fifth to the tobacco crop, would make the required increase.

The failure of the cotton crop, during the past season, (1866,) might have been entirely prevented, in many districts, by the thorough draining of the land.

The advantages claimed for drainage with reference to the above-named staple crops, will apply with equal, if not greater force, to all garden and orchard culture. In fact, with the exception of osier willows, and cranberries, there is scarcely a cultivated plant which will not yield larger and better crops on drained than on undrained land,—enough better, and enough larger, to pay much more than the interest on the cost of the improvement.

Yet, this advantage of draining, is, by no means, the only one which is worthy of consideration. Since the object of cultivation is to produce remunerative crops, of course, the larger and better the crops, the more completely is the object attained;—and to this extent the greatest benefit resulting from draining, lies in the increased yield. But there is another advantage,—a material and moral advantage,—which is equally to be considered.

Instances of the profit resulting from under-draining, (coupled, as it almost always is, with improved cultivation,) are frequently published, and it would be easy to fortify this chapter with hundreds of well authenticated cases. It is, however, deemed sufficient to quote the following, from an old number of one of the New York dailies: —

"Some years ago, the son of an English farmer came to the United States, and let himself as a farm laborer, in New York State, on the following conditions: Commencing work at the first of September, he was to work ten hours a day for three years, and to receive in payment a deed of a field containing twelve acres—securing himself by an agreement, by which his employer was put under bonds of $2,000 to fulfill his part of the contract; also, during these three years, he was to have the control of the field; to work it at his own expense, and to give his employer one-half the proceeds. The field lay under the south side of a hill, was of dark, heavy clay resting on a bluish-colored, solid clay subsoil, and for many years previous, had not been known to yield anything but a yellowish, hard, stunted vegetation.

"The farmer thought the young man was a simpleton, and that he, himself, was most wise and fortunate; but the former, nothing daunted by this opinion, which he was not unconscious that the latter entertained of him, immediately hired a set of laborers, and set them to work in the field trenching, as earnestly as it was well possible for men to labor. In the morning and evening, before and after having worked his ten hours, as per agreement, he worked with them, and continued to work in this way until, about the middle of the following November, he had finished the laying of nearly 5,000 yards of good tile under-drains. He then had the field plowed deep and thoroughly, and the earth thrown up as much as possible into ridges, and thus let it remain during the winter. Next spring he had the field again plowed as before, then cross-plowed and thoroughly pulverized with a heavy harrow, then sowed it with oats and clover. The yield was excellent—nothing to be compared to it had ever before been seen upon that field. Next year it gave two crops of clover, of a rich dark green, and enormously heavy and luxuriant; and the year following, after being manured at an expense of some $7 an acre, nine acres of the field yielded 936 bushels of corn, and 25 wagon loads of pumpkins; while from the remaining three acres were taken 100 bushels of potatoes—the return of this crop being upwards of $1,200. The time had now come for the field to fall into the young man's possession, and the farmer unhesitatingly offered him $1,500 to relinquish his title to it; and when this was unhesitatingly refused, he offered $2,000, which was accepted.

"The young man's account stood thus

Half proceeds of oats and straw, first year		$165 00
Half value of sheep pasturage, first year		25 00
Half of first crops of clover, first year		112 50
Half of second crops of clover, including seed, second year		135 00
Half of sheep pasturage, second year		15 00
Half of crops of corn, pumpkins and potatoes, third year		690 00
Received from farmer, for relinquishment of title		2,000 00
Account Dr.		$3,142 50
To under-draining, labor and tiles	$325 00	
To labor and manure, three seasons	475 00	
To labor given to farmer, $16 per month, 36 months	576 00	—1,376 00
Balance in his favor		$1,766 50

Draining makes the farmer, to a great extent, the master of his vocation. With a sloppy, drenched, cold, uncongenial soil, which is saturated with every rain, and takes days, and even weeks, to become sufficiently dry to work upon, his efforts are constantly baffled by unfavorable weather, at those times when it is most important that his work proceed without interruption. Weeks are lost, at a season when they are all too short for the work to be done. The ground must be hurriedly, and imperfectly prepared, and the seed is put in too late, often to rot in the over-soaked soil, requiring the field to be planted again at a time which makes it extremely doubtful whether the crop will ripen before the frost destroys it.

The necessary summer cultivation, between the rows, has to be done as the weather permits; and much more of it is required because of the baking of the ground. The whole life of the farmer, in fact, becomes a constant struggle with nature, and he fights always at a disadvantage. What he does by the work of days, is mainly undone by a single night's storm. Weeds grow apace, and the land is too wet to admit of their being exterminated. By the time that it is dry enough, other pressing work occupies the time; and if, finally, a day comes when they may be attacked, they offer ten times the resistance that they would have done a week earlier. The operations of the farm are carried on more expensively than if the ability to work constantly allowed a smaller force to be employed. The crops which give such doubtful promise, require the same cultivation as though they were certain to be remunerative, and the work can be done only with increased labor, because of the bad condition of the soil.

From force of tradition and of habit, the farmer accepts his fate and plods through his hard life, piously ascribing to the especial interference of an inscrutable Providence, the trials which come of his own neglect to use the means of relief which Providence has placed within his reach.

Trouble enough he must have, at any rate, but not necessarily all that he now has. It is not within the scope of the best laid drains to control storm or sunshine,—but it is within their power to remove the water of the storm, rapidly and sufficiently, and to allow the heat of the sunshine to penetrate the soil and do its hidden work. No human improvement can change any of the so-called "phenomena" of nature, or prevent the action of the least of her laws; but their effects upon the soil and its crops may be greatly modified, and that which, under certain circumstances, would have caused inconvenience or loss, may, by a change of circumstances, be made positively beneficial.

In the practice of agriculture, which is pre-eminently an economic art, draining will be prosecuted because of the pecuniary profit which it promises, and,—very properly,—it will not be pursued, to any considerable extent, where the money, which it costs, will not bring money in return. Yet, in a larger view of the case, its collateral advantages are of even greater moment than its mere profits. It is the foundation and the commencement of the most intelligent farming. It opens the way for other improvements, which, without it, would produce only doubtful or temporary benefits; and it enables the farmer so to extend and enlarge his operations, with fair promise of success, as to raise his occupation from a mere waiting upon the uncertain favors of nature, to an intelligent handling of her forces, for the attainment of almost certain results.

The rude work of an unthinking farmer, who scratches the surface soil with his plow, plants his seed, and trusts to the chances of a greater or less return, is unmitigated drudgery,—unworthy of an intelligent man; but he who investigates all of the causes of success and failure in farming, and adapts every operation to the requirements of the circumstances under which he works; doing everything in his power that may tend to the production of the results which he desires, and, so far as possible, avoiding everything that may interfere with his success,—leaving nothing to chance that can be secured, and securing all that chance may offer,—is engaged in the most ennobling, the most intelligent and the most progressive of all industrial avocations.

In the cultivation of retentive soils, drainage is the key to all improvement, and its advantage is to be measured not simply by the effect which it directly produces in increasing production, but, in still greater degree, by the extent to which it prepares the way for the successful application of improved processes, makes the farmer independent of weather and season, and offers freer scope to intelligence in the direction of his affairs.

CHAPTER VIII.

HOW TO MAKE DRAINING TILES.

Draining tiles are made of burnt clay, like bricks and earthen-ware.

In general terms, the process is as follows:—The clay is mixed with sand, or other substances which give it the proper consistency, and is so wetted as to form a plastic mass, to which may be given any desired form, and which is sufficiently stiff to retain its shape. Properly prepared clay is forced through the aperture of a die of the shape of the outside of the tile, while a plug,—held by a support in the rear of the die,—projects through the aperture, and gives the form to the bore of the tile. The shape of the material of the tile, as it comes from the die, corresponds to the open space, between the plug and the edge of the aperture. The clay is forced out in a continuous pipe, which is cut to the desired length by a wire, which is so thin as to pass through the mass without altering the shape of the pipe. The short lengths of pipe are dried in the air as thoroughly as they can be, and are then burned in a kiln, similar to that used for pottery.

Materials.—The range of earths which may be used in the manufacture of tiles is considerable, though clay is the basis of all of them. The best is, probably, the clay which is almost invariably found at the bottom of muck beds, as this is finer and more compact than that which is dug from dry land, and requires but little preparation. There is, also, a peculiar clay, found in some localities, which is almost like quick-sand in its nature, and which is excellent for tile-making,—requiring no freezing, or washing to prepare it for the machine. As a general rule, any clay which will make *good* bricks will make tiles. When first taken from the ground, these clays are not usually adhesive, but become so on being moistened and kneaded.

It is especially important that no limestone *pebbles* be mixed with the clay, as the burning would change these to quicklime, which, in slaking, would destroy the tiles. The presence of a limey earth, however, mixed through the mass, is a positive advantage, as in this intimate admixture, the lime forms, under the heat of the kiln, a chemical combination with the other ingredients; and, as it melts more readily than some of them, it hastens the burning and makes it more complete. What is known as *plastic clay*, (one of the purest of the native clays,) is too strong for tile-making, and must be "tempered," by having other substances mixed with it, to give it a stiffer quality.

The clay which is best for brick-making, contains Silica, and Alumina in about the following proportions:

Silica	55 to 75 per cent.
Alumina	35 " 25 " "

Variable quantities of other materials are usually found in connection with the clay, in its native condition. The most common of these are the following:—

Magnesia	1 to 5 per cent.—sometimes 20 to 30 per cent.
Lime	0 " 19 " "
Potash	0 " 5 " "
Oxyd of Iron	0 " 19 " "

"These necessary elements give fusibility to earthen-

"ware, and, therefore, allow its constituent substances to "combine in such a manner as to form a resisting body; "and this is performed with a temperature lower in pro- "portion as the necessary elements are more abundant."*

When the earth of the locality where tiles are to be made is not sufficiently strong for the purpose, and plastic clay can be cheaply obtained from a distance, a small quantity of this may be used to give strength and tenacity to the native material.

The compound must always contain a proper proportion of clay and sand. If too little *clay* is used, the mass will not be sufficiently tough to retain its compactness as it passes through the die of the tile machine; if too little *sand*, the moulded tiles will not be strong enough to bear handling, and they will crack and warp in drying and burning. Within the proper limits, the richer earths may be moulded much thinner, and tiles made from them may, consequently, be made lighter for transportation, without being too weak. The best materials for tempering stiff clays are sand, pounded brick or tile, or *scoria*, from smelting furnaces.

Preparation of Earths.—The clay from which tiles are to be made, should be thrown out in the fall, (the upper and lower parts of the beds being well mixed in the operation,) and made into heaps on the surface, not more than about 3 feet square and 3 feet high. In this form, it is left exposed to the freezing and thawing of winter, which will aid very much in modifying its character,—making it less lumpy and more easily workable. Any stones which may appear in the digging, should, of course, be removed, and most earths will be improved by being passed through a pair of heavy iron rollers, before they are piled up for the winter. The rollers should be made of cast iron, about 15 inches in diameter, and 30 inches long, and set as close

* Klippart's Land Drainage.

together as they can be, and still be revolved by the power of two horses. The grinding, by means of these rollers, may add 50 cents per thousand to the cost of the tiles, but it will greatly improve their quality.

In the spring, the clay should be prepared for tempering, by the removal of such pebbles as it may still contain. The best way to do this is by " washing," though, if there be only a few coarse pebbles, they may be removed by building the clay into a solid cone 2 or 3 feet high, and then paring it off into thin slices with a long knife having a handle at each end. This paring will discover any pebbles larger than a pea that may have remained in the clay.

Washing is the process of mixing the clay with a considerable quantity of water, so as to form a thin paste, in which all stones and gravel will sink to the bottom; the liquid portion is then drawn off into shallow pits or vats, and allowed to settle, the clear water being finally removed by pumping or by evaporation, according to the need for haste. For washing small quantities of clay, a common mortar bed, such as is used by masons, will answer, if it be supplied with a gate for draining off the muddy water after the gravel has settled; but, if the work is at all extensive, a washing mill will be required. It may be made in the form of a circular trough, with scrapers for mixing the clay and water attached to a circular horse-sweep.

"Another convenient mixing machine may be constructed " in the following manner: Take a large hollow log, of suit- " able length, say five or six feet; hew out the inequalities " with an adz, and close up the ends with pieces of strong " plank, into which bearing have been cut to support a re- " volving shaft. This shaft should be sufficiently thick to " permit being transfixed with wooden pins long enough to " reach within an inch or two of the sides of the log or " trough, and they should be so beveled as to form in their " aggregate shape an interrupted screw, having a direction "toward that end of the box where the mixed clay is de- "signed to pass out. In order to effect the mixing more "thoroughly, these pins may be placed sufficiently far apart "to permit the interior of the box to be armed with other "pins extending toward the center, between which they "can easily move. The whole is placed either horizontally "or vertically, and supplied with clay and water in proper "quantities, while the shaft is made to revolve by means of "a sweep, with horse power, running water or steam, as "the case may be. The clay is put into the end farthest "from the outlet, and is carried forward to it and mixed "by the motion, and mutual action and re-action of the pins "in the shaft and in the sides of the box. Iron pins may, "of course, be substituted for the wooden ones, and have "the advantage of greater durability and of greater strength "in proportion to their size, and the number may therefore "be greater in a machine of any given length. The fluid "mass of clay and water may be permitted to fall upon a "sieve or riddle, of heavy wire, and afterward be received "in a settling vat, of suitable size and construction, to drain "off the water and let the clay dry out sufficiently by sub- "sequent evaporation. A machine of this construction "may be made of such a size that it may be put in motion "by hand, by means of a crank, and yet be capable of "mixing, if properly supplied, clay enough to mold 800 "or 1000 pieces of drain pipe per day."*

Mr. Parkes, in a report to the Royal Agricultural Society of England, in 1843, says:

"It is requisite that the clay be well washed and sieved "before pugging, for the manufacture of these tiles, or the "operation of drawing them would be greatly impeded, by "having to remove stones from the small space surround- "ing the die, which determines the thickness of the pipe. "But it results from this necessary washing, that the sub-

* Klippart's Land Drainage.

"stance of the pipe is uniformly and extremely dense, "which, consequently, gives it immense strength, and en- "sures a durability which cannot belong to a more por- "ous, though thicker, tile.

"The clay is brought from the pug-mill so dry that, "when squeezed through the machine, not a drop of water "exudes,—moisture is, indeed, scarcely apparent on the "surface of the raw pipe. Hence, the tiles undergo little "or no change of figure while drying, which takes place "very rapidly, because of their firm and slight substance."

Tempering.—After the fine clay is relieved of the water with which it was washed, and has become tolerably dry, it should be mixed with the sand, or other tempering material, and passed through the *Pug-Mill*, (Fig. 42,) which will thoroughly mix its various ingredients, and work the whole into a homogeneous mass, ready for the tile machine. The *pug-mill* is similar to that used in brick-yards, only, as the clay is worked much stiffer for tiles than for bricks, iron knives must be substituted for the wooden pins. These knives are so arranged as to cut the clay in every part, and, by being set at an angle, they force it downward toward the outlet gate at the bottom. The clay should be kept at the proper degree of moisture from the time of tempering, and after passing through the pug-mill it should be thoroughly beaten to drive out the air, and the beaten mass should be kept covered with wet cloths to prevent drying.

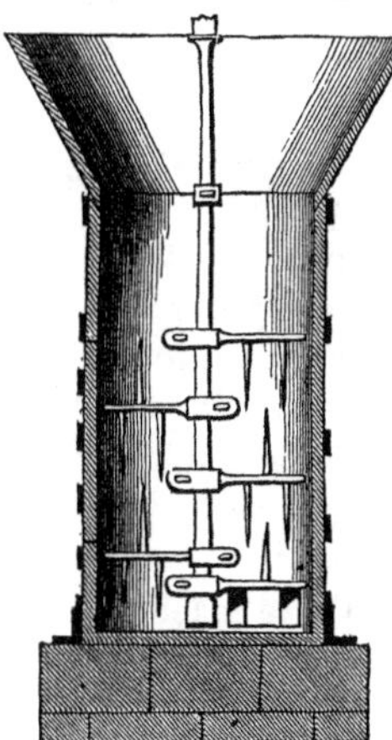
Fig. 42.—PUG-MILL.

Moulding the Tiles.—Machines for moulding tiles are

of various styles, with much variation in the details of their construction, but they all act on the same general principle;—that of forcing the clay through a ring-shaped aperture in an iron plate, forming a continuous pipe, which is carried off on an endless apron, or on rollers, and cut by wires into the desired lengths. The plates with the ring-shaped apertures are called *dies;* the openings are of any desired form, corresponding to the external shape of the tiles; and the size and shape of the bore, is determined by the core or plug, which is held in the centers of the apertures. The construction of the die plates, and the manner of fastening the plugs, which determine the bore of the tiles, is shown in Fig. 43. The view taken is of the inside of the plate.

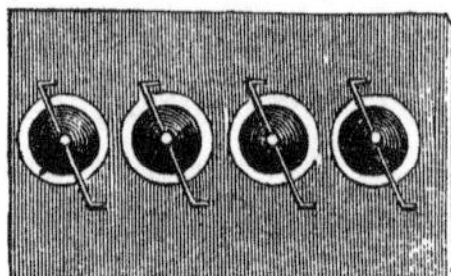
Fig. 43.—PLATE OF DIES.

The machine consists usually of a strong iron chest, with a hinged cover, into which the clay is placed, having a piston moving in it, connected by a rod or bar, having cog-teeth, with a cog-wheel, which is moved by horse or hand power, and drives the piston forward with steadiness, forcing the clay through the openings in the die-plate. The clay issues in continuous lines of pipe. The machines most in use in this country are connected directly with the pug-mill, and as the clay is pugged, it at once passes into the box, and is pressed out as tiles. These machines are usually run by horse-power.

Mr. Barral, in his voluminous work on drainage,* describes, as follows, a cheap hand machine which can be made by any country wheelwright, and which has a capacity of 3,000 tiles per day (Fig. 44):

" Imagine a simple, wooden box, divided into two com-
" partments. In the rear compartment there stands a
" vertical post, fastened with two iron bolts, having heads

*Drainage des Terres Arables, Paris, 1856.

" at one end, and nuts and screws at the other. The box " is thus fixed to its support. We simply place this sup- " port on the ground and bind its upper part with a rope " to a tree, a stake, or a post. The front compartment is " the reservoir for the clay, presenting at its front an " orifice, in which we fix the desired die with a simple bolt.

Fig. 44.—CHEAP WOODEN MACHINE.

" A wooden piston, of which the rod is jointed with a " lever, which works in a bolt at the top of the supporting " post, gives the necessary pressure. When the chest is " full of clay, we bear down on the end of the lever, " and the moulded tiles run out on a table supplied with " rollers. Raising the piston, it comes out of the box, " which is again packed with clay. The piston is replaced " in the box; pressure is again applied to the lever, and " so on. When the line of tiles reaches the end of the " table, we lower a frame on which brass wires are " stretched, and cut it into the usual lengths."

The workmen must attend well to the degree of moisture of the clay which is put into the machine. It should be dry enough to show no undue moisture on its surface as it comes out of the die-plate, and sufficiently moist not to be crumbled in passing the edge of the mould. The clay for small (thin) tiles must, necessarily, be more moist than that which is to pass through a wider aperture; and for the latter there may, with advantage, be more sand in the paste than would be practicable with the former.

After the tiles are cut into lengths, they are removed by a set of mandrils, small enough to pass easily into them, such as are shown in Fig. 45, (the number of fingers corresponding with the number of rows of tiles made by the machine,) and are placed on shelves made of narrow strips sawn from one-inch boards, laid with spaces between them to allow a free circulation of air.

Fig. 45.—MANDRIL FOR CARRYING TILES FROM MACHINE.

Drying and Rolling.—Care must be taken that freshly made tiles be not dried too rapidly. They should be sheltered from the sun and from strong winds. Too rapid drying has the effect of warping them out of shape, and, sometimes, of cracking the clay. To provide against this injury, the drying is done under sheds or other covering, and the side which is exposed to the prevailing winds is sometimes boarded up.

For the first drying, the tiles are placed in single layers on the shelves. When about half dried,—at which time they are usually warped more or less from their true shape,—it is well to *roll* them. This is done by passing through them a smooth, round stick, (sufficiently smaller than the bore to enter it easily, and long enough to project five or six inches beyond each end of the tile,) and,—holding one end of the stick in each hand,—rolling them carefully on a table. This operation should be performed when the tiles are still moist enough not to be broken by the slight bending required to make them straight. After rolling, the tiles may be piled up in close layers, some four or five feet high, (which will secure them against further warping,) and left until they are dry enough for burning,—that is, as dry as they can be made by exposure to the air.

Burning.—Tiles are burned in kilns in which, by the effect of flame acting directly upon them, they are raised to a heat sufficient to melt some of their more easily fusible ingredients, and give to them a stone-like hardness.

Kilns are of various construction and of various sizes. As this book is not intended for the instruction of those who are engaged in the general manufacture of tiles, only for those who may find it necessary to establish local works, it will be sufficient to describe a temporary earthen kiln which may be cheaply built, and which will answer an excellent purpose, where only 100,000 or 200,000 tiles per season will be required.

Directions for its construction are set forth in a letter from Mr. T. Law Hodges, of England, to the late Earl Spencer, published in the Journal of the Royal Agricultural Society for the year 1843, as follows:

"The form of the clay-kiln is circular, 11 feet in diame- "ter, and 7 feet high. It is wholly built of damp, clayey "earth, rammed firmly together, and plastered, inside and "out, with loam (clay ?). The earth to form the walls is dug "out around the base, leaving a circular trench about four "feet wide and as many deep, into which the fire-holes of "the kiln open. If wood be the fuel used, three fire-holes "will be sufficient; if coal, four will be needed. About "1,200 common brick will be wanted to build these fire- "holes and flues; if coal is used, rather fewer bricks will "be wanted, but, then, some iron bars are necessary,— "six bars to each fire-hole.

"The earthen walls are four feet thick at the floor of "the kiln, seven feet high, and tapering to a thickness of "two feet at the top; this will determine the slope of the "exterior face of the kiln. The inside of the wall is carried up perpendicularly, and the loam plastering inside becomes, after the first burning, like a brick wall. The kiln may be safely erected in March, or whenever the danger of injury from frost is over. After the summer use of it, it must be protected, by faggots or litter, against the wet and frost of winter. A kiln of these dimensions will contain 32,500 1¼-inch tiles, * * * or 12,000 2¼-inch tiles. * * *

"In good weather, this kiln can be filled, burnt, and discharged once in every fortnight, and fifteen kilns may be obtained in a good season, producing 487,500 1¼-inch tiles, and in proportion for the other sizes.

"It requires 2 tons 5 cwt. of good coals to burn the above kiln, full of tiles."

A sectional view of this kiln is shown in Fig. 46, in which *C*, *C* represent sections of the outer trench; *A*, one

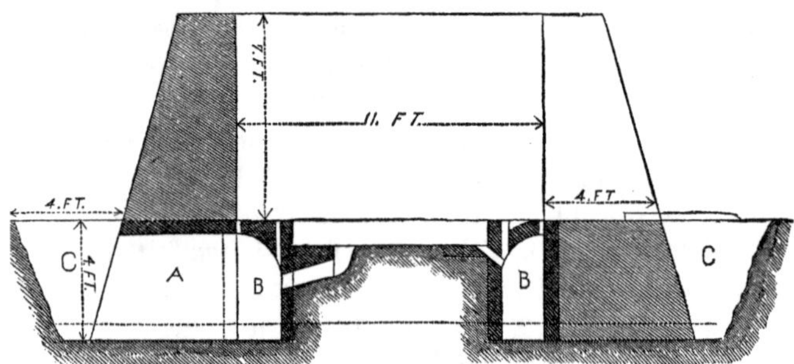

Fig. 46.—CLAY-KILN.

of the three fire-holes; and *B*, *B*, sections of a circular passage inside of the wall, connected with the fire-holes, and serving as a flue for the flames, which, at suitable intervals, pass through openings into the floor of the kiln. The whole structure should be covered with a roof of rough boards, placed high enough to be out of the reach of the fire. A door in the side of the kiln serves for putting in and removing the tiles, and is built up, temporarily, with bricks or clay, during the burning. Mr. Hodges estimates the cost of this kiln, all complete, at less than $25. Concerning its value, he wrote another letter in 1848, from which the following is extracted:

"The experience of four years that have elasped since "my letter to the late Earl Spencer, published in the 5th "volume of the proceedings of the Royal Agricultural "Society, page 57, has thoroughly tested the merits of "the temporary clay-kilns for the burning of draining-"pipes described in that letter.

"I am well aware that there were persons, even among "those who came to see it, who pronounced at once upon "the construction and duration of the kiln as unworthy "of attention. How far their expectations have been real-"ized, and what value belongs to their judgment, the fol-"lowing short statement will exhibit:

"The kiln, in question, was constructed, in 1844, at a "cost of £5.

"It was used four times in that year, burning each "time between 18,000 and 19,000 draining pipes, of 1¾ "inches in diameter.

"In 1845, it was used nine times, or about once a fort-"night, burning each time the same quantity of nearly "19,000 pipes.

"In 1846, the same result.

"In 1847, it has been used twelve times, always burn-"ing the same quantity. In the course of the last year a "trifling repair in the bottom of the kiln, costing rather "less than 10 shillings, was necessary, and this is the only "cost for repair since its erection. It is now as good as "ever, and might be worked at least once a fortnight "through the ensuing season.

"The result of this experiment of four years shows not "only the practical value of this cheap kiln, but Mr. "Hatcher, who superintends the brick and tile-yard at Ben-

" enden, where this kiln stands, expresses himself strongly " in favor of this kiln, as always producing better and " more evenly burned pipes than either of his larger and " better built brick-kilns can do."

The floor of the kiln is first covered with bricks, placed on end, at a little distance from each other, so as to allow the fire to pass between them, and the tiles are placed *on end* on these. This position will afford the best draft for the flames. After the kiln is packed full, the door-way is built up, and a slow fire is started,—only enough at first to complete the drying of the tiles, and to do this so slowly as not to warp them out of shape. They will be thoroughly dry when the smoke from the top of the kiln loses its dark color and becomes transparent. When the fires are well started, the mouths of the fire-holes may be built up so as to leave only sufficient room to put in fresh fuel, and if the wind is high, the fire-holes, on the side against which it blows, should be sheltered by some sort of screen which will counteract its influence, and keep up an even heat on all sides.

The time required for burning will be from two days and a night to four days and four nights, according to the dryness of the tiles, the state of the weather, and the character of the fuel. The fires should be drawn when the tiles in the hottest part of the kiln are burned to a "ringing" hardness. By leaving two or three holes in the door-way, which can be stopped with loose brick, a rod may be run in, from time to time, to take out specimen tiles from the hottest part of the kiln, which shall have been so placed as to be easily removed. The best plan, however,—the only prudent plan, in fact,—will be to employ an intelligent man who is thoroughly experienced in the burning of brick and pottery, and whose judgment in the management of the fires, and in the cooling off of the kiln, will save much of the waste that would result from inexperienced management. After the burning is completed, from

40 to 60 hours must be allowed for the cooling of the kiln before it is opened. If the cold air is admitted while it is still very hot, the unequal contraction of the material will cause the tiles to crack, and a large portion of them may be destroyed.

If any of the tiles are too much burned, they will be melted, and may stick together, or, at least, have their shape destroyed. Those which are not sufficiently burned would not withstand the action of the water in the soil, and should not be used. For the first of these accidents there is no remedy; for the latter, reburning will be necessary, and under-done tiles may be left, (or replaced,) in the kiln in the position which they occupied at the first burning, and the second heat will probably prove sufficient. There is less danger of unequal burning in circular than in square kilns. Soft wood is better than hard, as making a better flame. It should be split fine, and well seasoned.

Arrangement of the Tilery.—Such a tilery as is described above should have a drying shed from 60 to 80 feet long, and from 12 to 18 feet wide. This shed may be built in the cheapest and roughest manner, the roof being covered with felting, thatch, or hemlock boards, as economy may suggest. It should have a tier of drying shelves, (made of slats rather than of boards,) running the whole length of each side. A narrow, wooden tram-way, down the middle, to carry a car, by which the green tiles may be taken from the machine to the shelves, and the dry ones from the shelves to the kiln, will greatly lessen the cost of handling.

The pug-mill and tile-machine, as well as the clay pit and the washing-mill, should be at one end of the shed, and the kiln at the other, so that, even in rainy weather, the work may proceed without interruption. A shed of the size named will be sufficient to dry as many tiles of assorted sizes as can be burned in the clay-kiln described above.

The Cost of Tiles.—It would be impossible, at any time, to say what should be the precise cost of tiles in a given locality, without knowing the prices of labor and fuel; and in the present unsettled condition of the currency, any estimate would necessarily be of little value. Mr. Parker's estimated the cost of inch pipes in England at 6*s.*, (about $1.50,) per thousand, when made on the estate where they were to be used, by a process similar to that described herein. Probably they could at no time have been made for less than twice that cost in the United States, —and they would now cost much more; though if the clay is dug out in the fall, when the regularly employed farm hands are short of work, and if the same men can cut and haul the wood during the winter, the hands hired especially for the tile making, during the summer season, (two men and two or three boys,) cannot, even at present rates of wages, bring the cost of the tiles to nearly the market prices. If there be only temporary use for the machinery, it may be sold, when no longer needed, for a good percentage of its original cost, as, from the slow movement to which it is subjected, it is not much worn by its work.

There is no reason why tiles should cost more to make than bricks. A common brick contains clay enough to make four or five 1¼-inch tiles, and it will require about the same amount of fuel to burn this clay in one form as in the other. This advantage in favor of tiles is in a measure offset by the greater cost of handling them, and the greater liability to breakage.

The foregoing description of the different processes of the manufacture of draining tiles has been given, in order that those who find it necessary, or desirable, to establish works to supply the needs of their immediate localities may commence their operations understandingly, and form an approximate opinion of the promise of success in the undertaking.

Probably the most positive effect of the foregoing description, on the mind of any man who contemplates establishing a tilery, will be to cause him to visit some successful manufactory, during the busy season, and examine for himself the mode of operation. Certainly it would be unwise, when such a personal examination of the process is practicable, to rely entirely upon the aid of written descriptions; for, in any work like tile-making, where the selection, combination and preparation of the materials, the means of drying, and the economy and success of the burning must depend on a variety of conditions and circumstances, which change with every change of locality, it is impossible that written directions, however minute, should be a sufficient guide. Still, in the light of such directions, one can form a much better idea of the bearing of the different operations which he may witness, than he could possibly do if the whole process were new to him.

If a personal examination of a successful tilery is impracticable, it will be necessary to employ a practical brick-maker, or potter, to direct the construction and operation of the works, and in any case, this course is advisable.

In any neighborhood where two or three hundred acres of land are to be drained, if suitable earths can be readily obtained, it will be cheaper to establish a tile-yard, than to haul the necessary tiles, in wagons, a distance of ten or twenty miles. Then again, the prices demanded by the few manufacturers, who now have almost a monopoly of the business, are exorbitantly high,—at least twice what it will cost to make the tiles at home, with the cheap works described above, so that if the cost of transportation on the quantity desired would be equal to the cost of establishing the works, there will be a decided profit in the home manufacture. Probably, also, a tile-yard, in a neighborhood where the general character of the soil is such as to require drainage, will be of value after the object for which it was made has been accomplished.

While setting forth the advantage to the farmer of everything which may protect him against monopolies, whether in the matter of draining-tile, or of any other needful accessory of his business, or which will enable him to procure supplies without a ruinous outlay for transportation, it is by no means intended that every man shall become his own tile-maker.

In this branch of manufacture, as in every other, organized industry will accomplish results to which individual labor can never attain. A hundred years ago, when our mill-made cloths came from England, and cost more than farmers could afford to pay, they wore homespun, which was neither so handsome nor so good as the imported article; but, since that time, the growing population and the greater demand have caused cloth mills to be built here, greater commercial facilities have placed foreign goods within easy reach, and the house loom has fallen into general disuse.

At present, the manufacture of draining-tiles is confined to a few, widely separated localities, and each manufacturer has, thus far, been able to fix his own scale of charges. These, and the cost of transportation to distant points, make it difficult, if not impossible, for many farmers to procure tiles at a cost low enough to justify their use. In such cases, small works, to supply local demand, may enable many persons to drain with tiles, who, otherwise, would find it impossible to procure them cheaply enough for economical use; and the extension of underdraining, causing a more general acquaintance with its advantages, would create a sufficient demand to induce an increase of the manufacture of tiles, and a consequent reduction of price.

CHAPTER IX.

THE RECLAIMING OF SALT MARSHES.

"Adjoining to it is Middle Moor, containing about 2,500 acres, spoken "of by Arthur Young as 'a watery desert,' growing sedge and rushes, "and inhabited by frogs and bitterns;—it is now fertile, well cultivated, "and profitable land."

The foregoing extract, from an account of the Drainage of the Fens on the eastern coast of England, is a text from which might be preached a sermon worthy of the attention of all who are interested in the vast areas of salt marsh which form so large a part of our Atlantic coast, from Maine to Florida.

Hundreds of thousands of acres that might be cheaply reclaimed, and made our most valuable and most salubrious lands, are abandoned to the inroads of the sea;—fruitful only in malaria and musquitoes,—always a dreary waste, and often a grave annoyance.

A single tract, over 20,000 acres in extent, the center of which is not seven miles from the heart of New York City, skirts the Hackensack River, in New Jersey, serving as a barrier to intercourse between the town and the country which lies beyond it, adding miles to the daily travel of the thousands whose business and pleasure require them to cross it, and constituting a nuisance and an eyesore to all who see it, or come near it. How long it will continue in this condition it is impossible to say, but the experience of other countries has proved that, for an expense of not more than fifty dollars per acre, this tract might be made better, for all purposes of cultivation, than the lands adjoining it, (many of which are worth, for market gardening, over one thousand dollars per acre,) and that it might afford profitable employment, and give homes, to all of the industrious poor of the city. The work of reclaiming it would be child's play, compared with the draining of the Harlaem Lake in Holland, where over 40,000 acres, submerged to an average depth of thirteen feet, have been pumped dry, and made to do their part toward the support of a dense population.

The Hackensack meadows are only a conspicuous example of what exists over a great extent of our whole seaboard;—virgin lands, replete with every element of fertility, capable of producing enough food for the support of millions of human beings, better located, for residence and for convenience to markets, than the prairies of the Western States,—all allowed to remain worse than useless; while the poorer uplands near them are, in many places, teeming with a population whose lives are endangered, and whose comfort is sadly interfered with by the insects and the miasma which the marsh produces.

The inherent wealth of the land is locked up, and all of its bad effects are produced, by the water with which it is constantly soaked or overflowed. Let the waters of the sea be excluded, and a proper outlet for the rain-fall and the upland wash be provided,—both of which objects may, in a great majority of cases, be economically accomplished,—and this land may become the garden of the continent. Its fertility will attract a population, (especially in the vicinity of large towns,) which could no where else live so well nor so easily.

The manner in which these salt marshes were formed may be understood from the following account of the

"Great Level of the Fens" of the eastern coast of England, which is copied, (as is the paragraph at the head of this chapter,) from the Prize Essay of Mr. John Algernon Clarke, written for the Royal Agricultural Society in 1846.

The process is not, of course, always the same, nor are the exact influences, which made the English Fens, generally, operating in precisely the same manner here, but the main principle is the same, and the lesson taught by the improvement of the Fens is perfectly applicable in our case.

"This great level extends itself into the six counties of "Cambridge, Lincoln, Huntington, Northampton, Suffolk "and Norfolk, being bounded by the highlands of each. "It is about seventy miles in length, and varies from "twenty to forty miles in breadth, having an area of more "than 680,000 acres. Through this vast extent of flat "country, there flow six large rivers, with their tributary "streams; namely, the Ouse, the Cam, the Nene, the Wel-"land, the Glen, and the Witham.

"These were, originally, natural channels for conveying "the upland waters to the sea, and whenever a heavier "downfall of rain than usual occurred, and the swollen "springs and rivulets caused the rivers to overflow, they "must necessarily have overflowed the land to a great ex-"tent."

"This, however, was not the principal cause of the in-"undation of the Fens: these rivers were not allowed a "free passage to the ocean, being thus made incapable of "carrying off even the ordinary amount of upland water "which, consequently, flowed over the land. The obstruc-"tion was two-fold; first, the outfalls became blocked up "by the deposits of silt from the sea waters, which ac-"cumulated to an amazing thickness. The well known "instances of boats found in 1635 eight feet below the "Wisbeck River, and the smith's forge and tools found at "Skirbeck Shoals, near Boston, buried with silt sixteen feet "deep, show what an astonishing quantity of sediment

9

"formerly choked up the mouths of these great rivers. "But the chief hindrance caused by the ocean, arose from "the tide rushing twice every day for a very great dis- "tance up these channels, driving back the fresh waters, "and overflowing with them, so that the whole level be- "came deluged with deep water, and was, in fact, one "great bay."

"In considering the state of this region as it first at- "tracted the enterprise of man to its improvement, we "are to conceive a vast, wild morass, with only small, de- "tached portions of cultivated soil, or islands, raised above "the general inundation; a most desolate picture when "contrasted with its present state of matchless fertility."

Salt marshes are formed of the silty deposits of rivers and of the sea. The former bring down vegetable mould and fine earth from the uplands, and the latter contribute sea weeds and grasses, sand and shells, and millions of animalculæ which, born for life in salt water only, die, and are deposited with the other matters, at those points where, from admixture with the fresh flow of the rivers, the water ceases to be suitable for their support. It is estimated that these animalculæ alone are the chief cause of the obstructions at the mouths of the rivers of Holland, which retard their flow, and cause them to spread over the flat country adjoining their banks. It is less important, however, for the purposes of this chapter, to consider the manner in which salt marshes are formed, than to discuss the means by which they may be reclaimed and made available for the uses of agriculture. The improvement may be conveniently considered under three heads:—

First—The exclusion of the sea water.

Second—The removal of the causes of inundation from the upland.

Third—The removal of the rain-fall and water of filtration.

The Exclusion of the Sea is of the first importance, because not only does it saturate the land with water,—but this water, being salt, renders it unfertile for the plants of ordinary cultivation, and causes it to produce others which are of little, or no value.

The only means by which the sea may be kept out is, by building such dykes or embankments as shut out the highest tides, and, on shores which are exposed to the action of the waves, will resist their force. Ordinarily, the best, because the cheapest, material of which these embankments can be made, is the soil of the marsh itself. This is rarely,—almost never,—a pure peat, such as is found in upland swamps; it contains a large proportion of sand, blue clay, muscle mud, or other earthy deposits, which give it great weight and tenacity, and render it excellent for forming the body of the dyke. On lands which are overflowed to a considerable extent at each high tide, (twice a day,) it will be necessary to adopt more expensive, and more effective measures, but on ordinary salt meadows, which are deeply covered only at the spring tides, (occurring every month,) the following plan will be found practical and economical.

Locating the line of the embankment far enough back from the edge of the meadow to leave an ample flat outside of it to break the force of the waves, if on the open coast, or to resist the inroads of the current if on the bank of an estuary or a river,—say from ten to one hundred yards, according to the danger of encroachment,—set a row of stakes parallel to the general direction of the shore, to mark the outside line of the base of the dyke. Stake out the inside line at such distance as will give a pitch or inclination to the slopes of one and a half to one on the outside, and of one to one on the inside, and will allow the necessary width at the top, which should be at least two feet higher than the level of the highest tide that is known ever to have occurred at that place. The width of the top should never be less than four feet, and in exposed localities it should be more. If a road will be needed around the land, it is best, if a heavy dyke is required, to make it wide enough to answer this purpose, with still wider places, at intervals, to allow vehicles to turn or to pass each other. Ordinarily, however, especially if there be a good stretch of flat meadow in front, the top of the dyke need not be more than four feet wide. Supposing such a dyke to be contemplated where the water has been known to rise two feet above the level of the meadows, requiring an embankment four feet high, it will be necessary to allow for the base a width of fourteen feet;—four feet for the width of the top, six feet for the reach of the front slope, (1½ to 1,) and four feet for the reach of the back slope, (1 to 1.)

Having staked out two parallel lines, fourteen feet apart, and erected, at intervals of twenty or thirty feet, frames made of rough strips of board of the exact shape of the section of the proposed embankment, the workmen may remove the sod to a depth of six inches, laying it all on the outside of the position of the proposed embankment. The sod from the line of the ditch, from which the earth for the embankment is to be taken, should also be removed and placed with the other. This ditch should be always *inside* of the dyke, where it will never be exposed to the action of the sea. It should be, at the surface, broader than the base of the dyke, and five feet deep in the center, but its sides may slope from the surface of the ground directly to the center line of the bottom. This is the best form to give it, because, while it should be five feet deep, for future uses as a drain, its bottom need have no width. The great width at the surface will give such a pitch to the banks as to ensure their stability, and will yield a large amount of sod for the facing of the dyke. The edge of this ditch should be some feet away from the inner line of the embankment, leaving it a firm support or shoulder at the original level of the ground, the sod not being removed from the interval. The next step in the work should be to throw, or wheel, the material from the ditch on to the place which has been stripped for the dyke, building it up so as to conform exactly to the profile frames, these remaining in their places, to indicate the filling necessary to make up for the settling of the material, as the water drains out of it.

As fast as a permanent shape can be given to the outer face of the dyke, it should be finished by having the sod placed against it, being laid flatwise, one on top of another, (like stone work,) in the most solid manner possible. This should be continued to the top of the slope, and the flat top of the dyke should also be sodded,—the sods on the top, and on the slope, being firmly beaten to their places with the back of the spade or other suitable implement.

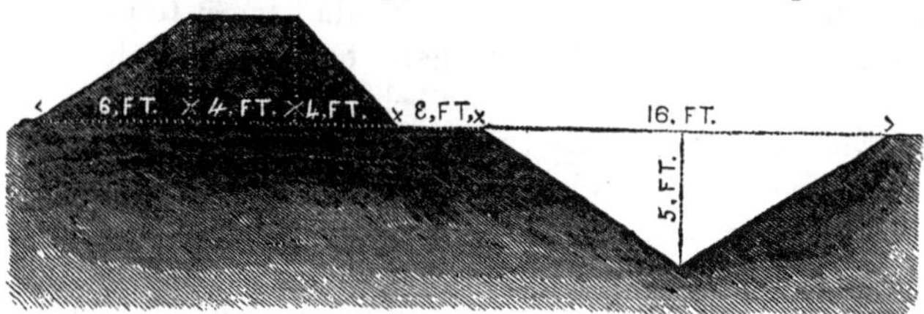

Fig. 47.—DYKE AND DITCH.

This will sufficiently protect the exposed parts of the work against the action of any waves that may be formed on the flat between the dyke and the deep water, while the inner slope and the banks of the ditch, not being exposed to masses of moving water, will retain their shape and will soon be covered with a new growth.* A sectional view of the above described dyke and ditch is shown in the accompanying diagram, (Fig. 47.)

* The ends of the work, while the operations are suspended during spring tides, will need an extra protection of sods, but that lying out of reach of the eddies that will be formed by the receding water will not be materially affected.

In all work of this character, it is important to regulate the amount of work laid out to be done between the spring tides, to the laboring force employed, so that no unfinished work will remain to be submerged and injured. When the flood comes, it should find everything finished up and protected against its ravages, so that no part of it need be done over again.

If the land is crossed by creeks, the dyke should be finished off and sodded, a little back from each bank, and when the time comes for closing the channel, sufficient force should be employed to complete the dam at a single tide, so that the returning flow shall not enter to wash away the material which has been thrown in.

If, as is often the case, these creeks are not merely tidal estuaries, but receive brooks or rivers from the upland, provision must be made, as will be hereafter directed, for either diverting the upland flow, or for allowing it to pass out at low water, through valve gates or sluices. When the dam has been made, the water behind it should never be allowed to rise to nearly the level of the full tide, and, as soon as possible, grass and willows should be grown on the bank, to add to its strength by the binding effect of their roots.

When the dyke is completed across the front of the whole flat,—from the high land on one side to the high land on the other, the creeks should be closed, one after the other, commencing with the smallest, so that the experience gained in their treatment may enable the force to work more advantageously on those which carry more water.

If the flow of water in the creek is considerable, a row of strong stakes, or piles, should be firmly driven into the bottom mud, across the whole width of the channel, at intervals of not more than one or two feet, and *fascines*,—bundles of brush bound together,—should be made ready on the banks, in sufficient quantity to close the spaces between the piles. These will serve to prevent the washing away of the filling during construction. The pile driving, and the preparation of the fascines may be done before the closing of the channel with earth is commenced, and if upland clay or gravel, to be mixed with the local material, can be economically brought to the place by boats or wagons, it will be an advantage. Everything being in readiness, a sufficient force of laborers to finish the dam in six hours should commence the work a little before dead low-water, and, (with the aid of wheelbarrows, if necessary,) throw the earth in rapidly *behind* the row of stakes and fascines, giving the dam sufficient width to resist the pressure of the water from without, and keeping the work always in advance of the rising of the tide, so that, during the whole operation, none of the filling shall be washed away by water flowing over its top.

If the creek has a sloping bottom, the work may be commenced earlier,—as soon as the tide commences to recede,—and pushed out to the center of the channel by the time the tide is out. When the dam is built, it will be best to heavily sod, or otherwise protect its surface against the action of heavy rains, which would tend to wash it away and weaken it; and the bed of the creek should be filled in back of the dam for a distance of at least fifty yards, to a height greater than that at which water will stand in the interior drains,—say to within three feet of the surface,—so that there shall never be a body of water standing within that distance of the dam.

This is a necessary precaution against the attacks of muskrats, which are the principal cause of the insecurity of all salt marsh embankments. It should be a cardinal rule with all who are engaged in the construction of such works, never to allow two bodies of water, one on each side of the bank to be nearer than twenty-five yards of each other, and fifty yards would be better. Muskrats do not bore through a bank, as is often supposed, to make a passage from one body of water to another, (they would find an easier road over the top); but they delight in any elevated mound in which they can make their homes above the water level and have its entrance beneath the surface, so that their land enemies cannot invade them. When they enter for this purpose, only from one side of the dyke, they will do no harm, but if another colony is, at the same time, boring in from the other side, there is great danger that their burrows will connect, and thus form a channel for the admission of water, and destroy the work. A disregard of this requirement has caused thousands of acres of salt marsh that had been enclosed by dykes having a ditch on each side, (much the cheapest way to make them,) to be abandoned, and it has induced the invention of various costly devices for the protection of embankments against these attacks.*

When the creek or estuary to be cut off is very wide, the embankment may be carried out, at leisure, from each side, until the channel is only wide enough to allow the passage of the tide without too great a rush of water against the unfinished ends of the work; but, even in these cases, there will be economy in the use of fascines and piles from the first, or of stones if these can be readily procured. In wide streams, partial obstructions of the water

* The latest invention of this sort, is that of a series of cast iron plates, set on edge, riveted together, and driven in to such a depth as to reach from the top of the dyke to a point below low-water mark. The best that can be said of this plan is, that its adoption would do no harm. Unless the plates are driven deeply into the clay underlying the permeable soil, (and this is sometimes very deep,) they would not prevent the slight infiltration of water which could pass under them as well as through any other part of the soil, and unless the iron were very thick, the corrosive action of salt water would soon so honeycomb it that the borers would easily penetrate it; but the great objection to the use of these plates is, that they would be very costly and ineffectual. A dyke, made as described above, of the material of the locality, having a ditch only on the inside, and being well sodded on its outer face, would be far cheaper and better.

course will sometimes induce the deposit of silt in such quantities as will greatly assist the work. No written description of a single process will suffice for the direction of those having charge of this most delicate of all drainage operations. Much must be left to the ingenuity of the director of the work, who will have to avail himself of the assistance of such favorable circumstances as may, in the case in hand, offer themselves.

If the barrier to be built will require a considerable outlay, it should be placed in the hands of a competent engineer, and it will generally demand the full measure of his skill and experience.

The work cannot be successful, unless the whole line of the water-front is protected by a continuous bank, sufficiently high and strong in all of its parts to resist the action of the highest tides and the strongest waves to which it will be subjected. As it is always open to inspection, at each ebb tide, and can always be approached for repair, it will be easy to keep it in good condition; and, if properly attended to, it will become more solid and effective with age.

The removal of the causes of inundation from the upland is often of almost equal importance with the shutting out of the sea, since the amount of water brought down by rivers, brooks, and hill-side wash, is often more than can be removed by any practicable means, by sluice gates, or pumps.

It will be quite enough for the capacity of these means of drainage, to remove the rain-water which falls on the flat land, and that which reaches it by under-ground springs and by infiltration,—its proper drainage-water in short,—without adding that which, coming from a higher level, may be made to flow off by its own fall.

Catch-water drains, near the foot of the upland, may be so arranged as to receive the surface water of the hills and carry it off, always on a level above that of the top of the embankment, and these drains may often be, with advantage, enlarged to a sufficient capacity to carry the streams as well. If the marsh is divided by an actual river, it may be best to embank it in two separate tracts; losing the margins, that have been recommended, outside of the dykes, and building the necessary additional length of these, rather than to contend with a large body of water. But, frequently, a very large marsh is traversed by a tortuous stream which occupies a large area, and which, although the tidal water which it contains gives it the appearance of a river, is only the outlet of an insignificant stream, which might be carried along the edge of the upland in an ordinary mill-race. In such case it is better to divert the stream and reclaim the whole area.

When a stream is enclosed between dykes, its winding course should be made straight in order that its water may be carried off as rapidly as possible, and the land which it occupies by its deviations, made available for cultivation. In the loose, silty soil of a salt marsh, the stream may be made to do most of the work of making its new bed, by constructing temporary "jetties," or other obstructions to its accustomed flow, which shall cause its current to deposit silt in its old channel, and to cut a new one out of the opposite bank. In some instances it may be well to make an elevated canal, straight across the tract, by constructing banks high enough to confine the stream and deliver it over the top of the dyke; in others it may be more expedient to carry the stream over, or through, the hill which bounds the marsh, and cause it to discharge through an adjoining valley. Improvements of this magnitude, which often affect the interest of many owners, or of persons interested in the navigation of the old channel, or in mill privileges below the point at which the water course is to be diverted, will generally require legislative interference.

But they not seldom promise immense advantages for a comparatively small outlay.

The instance cited of the Hackensack Meadows, in New Jersey, is a case in point. Its area is divided among many owners, and, while ninety-nine acres in every hundred are given up to muskrats, mosquitoes, coarse rushes and malaria, the other one acre may belong to the owner of an adjacent farm who values the salt hay which it yields him, and the title to the whole is vested in many individual proprietors, who could never be induced to unite in an improvement for the common benefit. Then again, thanks to the tide that sets back in the Hackensack River, it is able to float an occasional vessel to the unimportant villages at the northern end of the meadows, and the right of navigation can be interfered with only by govermental action. If the Hackensack River proper, that part of it which only serves as an outlet for the drainage of the high land north of the meadows, could be diverted and carried through the hills to the Passaic; or confined within straight elevated banks and made to discharge at high water mark at the line of the Philadelphia Rail-road ; —the wash of the highlands, east and west of the meadows, being also carried off at this level,—the bridge of the railroad might be replaced by an earth embankment, less than a quarter of a mile in length, effecting a complete exclusion of the tidal flow from the whole tract.

This being done, a steam-pump, far less formidable than many which are in profitable use in Europe for the same purpose, would empty, and keep empty, the present bed of the river, which would form a capital outlet for the drainage of the whole area. Twenty thousand acres, of the most fertile land, would thus be added to the available area of the State, greatly increasing its wealth, and inducing the settlement of thousands of industrious inhabitants.

As the circumstances under which upland water reaches

lands of the class under consideration vary with every locality, no specific directions for the treatment of individual cases can be given within the limits of this chapter; but the problem will rarely be a difficult one.

The removal of the rain-fall and water of filtration is the next point to be considered.

So far as the drainage of the land, in detail, is concerned, it is only necessary to say that it may be accomplished, as in the case of any other level land which, from the slight fall that can be allowed the drains, requires close attention and great care in the adjustment of the grades.

The main difficulty is in providing an outlet for the drains. This can only be done by artificial means, as the water must be removed from a level lower than high-water mark,—sometimes lower than low-water.

If it is only required that the outlet be at a point somewhat above the level of ordinary low-water, it will be sufficient to provide a sufficient reservoir, (usually a large open ditch,) to contain the drainage water that is discharged while the tide stands above the floor of the outlet sluice-way, and to provide for its outflow while the level of the tide water is below the point of discharge. This is done by means of sluices having self-acting valves, (or tide-gates,) opening outward, which will be closed by the weight of the water when the tide rises against them, being opened again by the pressure of the water from within, as soon the tide falls below the level of the water inside of the bank.

The gates and sluices may be of wood or iron,—square or round. The best would be galvanized iron pipes and valves; but a square wooden trunk, closed with a heavy oak gate that fits closely against its outer end, and moves freely on its hinges, will answer capitally well, if carefully and strongly made. If the gate is of wood, it will be well to have it lie in a slightly slanting position, so that its own weight will tend to keep it closed when the tide first commences to rise above the floor, and might trickle in, before it had acquired sufficient head to press the gate against the end of the trunk.

As this outlet has to remove, in a short time, all of the water that is delivered by the drains and ditches during several hours, it should, of course, be considerably larger than would be required for a constantly flowing drain from the same area; but the immense gates,—large enough for a canal lock,—which are sometimes used for the drainage of a few acres of marsh, are absurd. Not only are they useless, they are really objectionable, inasmuch as the greater extent of their joints increases the risk of leakage at the time of high water.

The channel for the outflow of the water may sometimes, with advantage, be open to the top of the dyke or dam,—a canal instead of a trunk; but this is rarely the better plan, and is only admissible where the discharge is into a river or small bay, too small for the formation of high waves, as these would be best received on the face of a well sodded, sloping bank.

The height, above absolute low water, at which the outlet should be placed, will depend on the depth of the outlet of the land drain, and the depth of storage room required to receive the drainage water during the higher stages of the tide. Of course, it must not be higher than the floor of the land drain outlet, and, except for the purpose of affording storage room, it need not be lower, although all the drainage will discharge, not only while the tide water is below the bottom of the gate, but as long as it remains lower than the level of the water inside. It is well to place the mouth of the trunk nearly as low as ordinary low-water mark. This will frequently render it necessary to carry a covered drain, of wood or brick, through the mud, out as far as the tide usually recedes,—connected with the valve gate at the outlet of the trunk, by a covered box which will keep rubbish from obstructing it, or interfering with its action.

When the outlet of the land-drains is below low-water mark, it is of course necessary to pump out the drainage water. This is done by steam or by wind, the latter being economical only for small tracts which will not bear the cost of a steam pump. Formerly, this work was done entirely by windmills, but these afford only an uncertain power, and often cause the entire loss of crops which are ready for the harvest, by obstinately refusing to work for days after a heavy rain has deluged the land. In grass land they are tolerably reliable, and on *small* tracts in cultivation, it is easy, by having a good proportion of open ditches, to afford storage room sufficient for general security; but in the reclaiming of large areas, (and it is with these that the work is most economical,) the steam pump may be regarded as indispensable. It is fast superseding the windmills which, a few years ago, were the sole dependence in Holland and on the English Fens. The magnitude of the pumping machinery on which the agriculture of a large part of Holland depends, is astonishing.

There are such immense areas of salt marsh in the United States which may be tolerably drained by the use of simple valve gates, discharging above low-water mark, that it is not very important to consider the question of pumping, except in cases where owners of small tracts, from which a sufficient tidal outlet could not be secured, (without the concurrence of adjoining proprietors who might refuse to unite in making the improvement,) may find it advisable to erect small pumps for their own use. In such cases, it would generally be most economical to use wind-power, especially if an accessory steam pump be provided for occasional use, in emergency. Certainly, the tidal drainage should first be resorted to, for when the land has once been brought into cultivation, the propriety of introducing steam pumps will become more apparent, and the outlay will be made with more confidence of profitable return, and, in all cases, the tidal outlet should be depended on for the outflow of all water above its level. It would be folly to raise water by expensive means, which can be removed, even periodically, by natural drainage.

When pumps are used, their discharge pipes should pass through the embankment, and deliver the water at low-water mark, so that the engine may have to operate only against the actual height of the tide water. If it delivered above high-water mark, it would work, even at low tide, against a constant head, equal to that of the highest tides.

CHAPTER X.

MALARIAL DISEASES.

So far as remote agricultural districts are concerned, it is not probable that the mere question of health would induce the undertaking of costly drainage operations, although this consideration may operate, in connection with the need for an improved condition of soil, as a strong argument in its favor. As a rule, " the chills " are accepted by farmers, especially at the West, as one of the slight inconveniences attending their residence on rich lands; and it is not proposed, in this work, to urge the evils of this terrible disease, and of "sun pain," or " day neuralgia," as a reason for draining the immense prairies over which they prevail. The diseases exist,—to the incalculable detriment of the people,—and thorough draining would remove them, and would doubtless bring a large average return on the investment; — but the question is, after all, one of capital; and the cost of such draining as would remove fever-and-ague from the bottom lands and prairies of the West, and from the infected agricultural districts at the East, would be more than the agricultural capital of those districts could spare for the purpose.

In the vicinity of cities and towns, however, where more wealth has accumulated, and where the number of persons subjected to the malarial influence is greater, there can be no question as to the propriety of draining, even if nothing but improved health be the object.

Then again, there are immense tracts near the large cities of this country which would be most desirable for residence, were it not that their occupancy, except with certain constant precautions, implies almost inevitable suffering from fever-and-ague, or neuralgia.

Very few neighborhoods within thirty miles of the city of New York are entirely free from these scourges, whose influence has greatly retarded their occupation by those who are seeking country homes; while many, who have braved the dangers of disease in these localities, have had sad cause to regret their temerity.

Probably the most striking instance of the effect of malaria on the growth and settlement of suburban districts, is to be found on Staten Island. Within five miles of the Battery; accessible by the most agreeable and best managed ferry from the city; practically, nearer to Wall street than Murray Hill is; with most charming views of land and water; with a beautifully diversified surface, and an excellent soil; and affording capital opportunities for sea bathing, it should be, (were it not for its sanitary reputation, it inevitably would be,) one vast residence-park. Except on its extreme northern end, and along its higher ridges, it has,—and, unfortunately, it deserves,—a most unenviable reputation for insalubrity. Here and there, on the southern slope also, there are favored places which are unaccountably free from the pest, but, as a rule, it is, during the summer and autumn, unsafe to live there without having constant recourse to preventive medication, or exercising unusual and inconvenient precautions with regard to exposure to mid-day sun and evening dew. There are always to be found attractive residences, which are deserted by their owners, and are offered for sale at absurdly low prices. There are isolated instances of very thorough and very costly draining, which has failed of effect, because so extensive a malarial region cannot be reclaimed by anything short of a systematic improvement of the whole.

It has been estimated that the thorough drainage of the low lands, valleys and ponds of the eastern end of the island, including two miles of the south shore, would at once add $5,000,000 to the market value of the real estate of that section. There can be no question that any radical improvement in this respect would remove the only obstacle to the rapid settlement of the island by those who wish to live in the country, yet need to be near to the business portion of the city. The hope of such improvement being made, however, seems as remote as ever,—although any one at all acquainted with the sources of miasm, in country neighborhoods, can readily see the cause of the difficulty, and the means for its removal are as plainly suggested.

Staten Island is, by no means, alone in this respect. All who know the history of the settlement of the other suburbs of New York are very well aware that those places which are free from fever-and-ague and malarial neuralgia, are extremely rare.

The exact cause of fever-and-ague and other malarial diseases is unknown, but it is demonstrated that, whatever the cause is, it is originated under a combination of circumstances, one of which is undue moisture in the soil. It is not necessary that land should be absolutely marshy to produce the miasm, for this often arises on cold, springy uplands which are quite free from deposits of muck. Thus far, the attention of scientific investigators, given to the consideration of the origin of malarial diseases, has failed to discover any well established facts concerning it; but there have been developed certain theories, which seem to be sustained by such knowledge as exists on the subject.

Dr. Bartlett, in his work on the Fevers of the United States, says: — "The essential, efficient, producing cause "of periodical fever, — the poison whose action on "the system gives rise to the disease,—is a substance or "agent which has received the names of *malaria*, or *marsh* "*miasm.* The nature and composition of this poison are "wholly unknown to us. Like most other analagous "agents, like the contagious principle of small-pox and of "typhus, and like the epidemic poison of scarletina and "cholera, they are too subtle to be recognized by any "of our senses, they are too fugitive to be caught by any "of our contrivances.

"As always happens in such cases and under similar "circumstances, in the absence of positive knowledge, we "have been abundantly supplied with conjecture and spec-"ulation; what observation has failed to discover, hy-"pothesis has endeavored and professed to supply. It is "quite unncessary even to enumerate the different sub-"stances to which malaria has been referred. Amongst "them are all of the chemical products and compounds "possible in wet and marshy localities; moisture alone; "the products of animal and vegetable decomposition; "and invisible living organisms. * * * * Inscruta-"ble, however, as the intimate nature of the substances "or agents may be, there are some few of its laws and "relations which are very well ascertained. One of these "consists in its connection with low, or wet, or marshy "localities. This connection is not invariable and exclu-"sive, that is, there are marshy localities which are not "malarious, and there are malarious localities which are "not marshy; but there is no doubt whatever that it gen-"erally exists."

In a report to the United States Sanitary Commission, Dr. Metcalfe states, that all hypotheses, even the most plausible, are entirely unsupported by positive knowledge, and he says: —

"This confession of ignorance still leaves us in posses- "sion of certain knowledge concerning malaria, from which "much practical good may be derived.

"1st. It affects, by preference, low and moist localities.

"2d. It is almost never developed at a lower tempera- "ture than 60° Fahrenheit.

"3d. Its evolution or active agency is checked by a "temperature of 32°.

"4th. It is most abundant and most virulent as we ap- "proach the equator and the sea-coast.

"5th. It has an affinity for dense foliage, which has the "power of accumulating it, when lying in the course of "winds blowing from malarious localities.

"6th. Forests, or even woods, have the power of ob- "structing and preventing its transmission, under these "circumstances.

"7th. By atmospheric currents it is capable of being "transported to considerable distances—probably as far as "five miles.

"8th. It may be developed, in previously healthy places, "by turning up the soil; as in making excavations for "foundations of houses, tracks for railroads, and beds for "canals.

"9th. In certain cases it seems to be attracted and ab- "sorbed by bodies of water lying in the course of such "winds as waft it from the miasmatic source.

"10th. Experience alone can enable us to decide as to "the presence or absence of malaria, in any given locality.

"11th. In proportion as countries, previously malarious, "are cleared up and thickly settled, periodical fevers dis- "appear—in many instances to be replaced by the typhoid "or typhus."

La Roche, in a carefully prepared treatise on "Pneumonia; its Supposed Connection with Autumnal Fevers," recites various theories concerning the mode of action of marsh miasm, and finds them insufficient to account for the phenomena which they produce. He continues as follows: —

"All the above hypotheses failing to account for the ef- "fects in question, we are naturally led to the admission "that they are produced by the morbific influence of some "special agent; and when we take into consideration all "the circumstances attending the appearance of febrile "diseases, the circumscribed sphere of their prevalence, "the suddenness of their attack, the character of their "phenomena, etc., we may safely say that there is noth- "ing left but to attribute them to the action of some "poison dissolved or suspended in the air of the infected "locality; which poison, while doubtless requiring for its "development and dissemination a certain degree of heat. "and terrestrial and atmospheric moisture, a certain "amount of nightly condensation after evaporation, and "the presence of fermenting or decomposing materials, "cannot be produced by either of these agencies alone, "and though indicated by the chemist, betrays its pres- "ence by producing on those exposed to its influence the "peculiar morbid changes characterizing fever."

He quotes the following from the Researches of Dr. Chadwick: —

"In considering the circumstances external to the resi- "dence, which affect the sanitary condition of the popula- "tion, the importance of a general land-drainage is devel- "oped by the inquiries as to the cause of the prevalent "diseases, to be of a magnitude of which no conception had "been formed at the commencement of the investigation. "Its importance is manifested by the severe consequences "of its neglect in every part of the country, as well as by "its advantages in the increasing salubrity and productive- "ness wherever the drainage has been skillful and effec- "tual."

La Roche calls attention to these facts:—That the acclimated residents of a malarious locality, while they are less subject than strangers to active fevér, show, in their physical and even in their mental organization, evident indications of the ill effects of living in a poisonous atmosphere,—an evil which increases with successive generations, often resulting in a positive deterioration of the race; that the lower animals are affected, though in a less degree than man; that deposits of organic matter which are entirely covered with water, (as at the bottom of a pond,) are not productive of malaria; that this condition of saturation is infinitely preferable to imperfect drainage; that swamps which are shaded from the sun's heat by trees, are not supposed to produce disease; and that marshes which are exposed to constant winds are not especially deleterious to persons living in their immediate vicinity,—while winds frequently carry the emanations of miasmatic districts to points some miles distant, where they produce their worst effects. This latter statement is substantiated by the fact that houses situated some miles to the leeward of low, wet lands, have been especially insalubrious until the windows and doors on the side toward the source of the miasm were closed up, and openings made on the other side,—and thenceforth remained free from the disease, although other houses with openings on the exposed sides continued unhealthy.

The literature relating to periodical fevers contains nothing else so interesting as the very ingenious article of Dr. J. H. Salisbury, on the "Cause of Malarious Fevers," contributed to the "American Journal of Medical Science," for January, 1866. Unfortunately, while there is no evidence to controvert the statements of this article, they do not seem to be honored with the confidence of the profession,—not being regarded as sufficiently authenticated to form a basis for scientific deductions. Dr. Salisbury claims to have discovered the cause of malarial fever in the spores of a very low order of plant, which spores he claims to have invariably detected in the saliva, and in the urine, of fever patients, and in those of no other persons, and which he collected on plates of glass suspended over all marshes and other lands of a malarious character, which he examined, and which he was never able to obtain from lands which were not malarious. Starting from this point, he proceeds, (with circumstantial statements that seem to the unprofessional mind to be sufficient,) to show that the plant producing these spores is always found, in the form of a whitish, green, or brick-colored incrustation, on the surface of fever producing lands; that the spores, when detached from the parent plant, are carried in suspension *only in the moist exhalations of wet lands*, never rising higher, (usually from 35 to 60 feet,) nor being carried farther, than the humid air itself; that they most accumulate in the upper strata of the fogs, producing more disease on lands slightly elevated above the level of the marsh than at its very edge; that fever-and-ague are never to be found where this plant does not grow; that it may be at once introduced into the healthiest locality by transporting moist earth on which the incrustation is forming; that the plant, being introduced into the human system through the lungs, continues to grow there and causes disease; and that *quinia* arrests its growth, (as it checks the multiplication of yeast plants in fermentation,) and thus suspends the action of the disease.

Probably it would be impossible to prove that the foregoing theory is correct, though it is not improbable that it contains the germ from which a fuller knowledge of the disease and its causes will be obtained. It is sufficient for the purposes of this work to say that, so far as Dr. Salisbury's opinion is valuable, it is,—like the opinion of all other writers on the subject,—fully in favor of perfect drainage as the one great preventive of all malarial diseases.

The evidence of the effect of drainage in removing the cause of malarial diseases is complete and conclusive. Instances of such improvement in this country are not rare, but they are much less numerous and less conspicuous here than in England, where draining has been much more extensively carried out, and where greater pains have been taken to collect testimony as to its effects.

If there is any fact well established by satisfactory experience, it is that thorough and judicious draining will entirely remove the local source of the miasm which produces these diseases.

The voluminous reports of various Committees of the English Parliament, appointed to investigate sanitary questions, are replete with information concerning experience throughout the whole country, bearing directly on this question.

Dr. Whitley, in his report to the Board of Health, (in 1864,) of an extended tour of observation, says of one town that he examined:—

"Mr. Nicholls, who has been forty years in practice "here, and whom I was unable to see at the time of my "visit, writes: Intermittent and remittent are greatly on "on the decline since the improved state of drainage of "the town and surrounding district, and more particularly "marked is this alteration, since the introduction of the "water-works in the place. Although we have occasional "outbreaks of intermittent and remittent, with neuralgic "attacks, they yield more speedily to remedies, and are "not attended by so much enlargement of the liver or "spleen as formerly, and dysentery is of rare occurrence."

Dr. Whitley sums up his case as follows:—

"It would appear from the foregoing inquiry, that in"termittent and remittent fevers, and their consequences, "can no longer be regarded as seriously affecting the "health of the population, in many of the districts, in which "those diseases were formerly of a formidable character.

"Thus, in Norfolk, Lincolnshire, and Cambridgeshire, " counties in which these diseases were both frequent and " severe, all the evidence, except that furnished by the " Peterborough Infirmary, and, in a somewhat less degree, " in Spaulding, tends to show that they are at the present " time, comparatively rare and mild in form."

* * * * * * * * *

He mentions similar results from his investigations in other parts of the kingdom, and says: —

" It may, therefore, be safely asserted as regards Eng- " land generally, that:—

" The diseases which have been made the subject of the " present inquiry, have been steadily decreasing, both in " frequency and severity, for several years, *and this de- " crease is attributed, in nearly every case, mainly to one " cause,—improved land drainage;*" again:

" The change of local circumstances, unanimously de- " clared to be the most immediate in influencing the pre- " valence of malarious diseases, is land drainage;" and again:

" Except in a few cases in which medical men believed " that these affections began to decline previously to the " improved drainage of the places mentioned, the decrease " in all of the districts where extensive drainage has been " carried out, was stated to have commenced about the " same time, and was unhesitatingly attributed to that " cause."

A select Committee of the House of Commons, appointed to investigate the condition and sanitary influence of the Thames marshes, reported their minutes of evidence, and their deductions therefrom, in 1854. The following is extracted from their report:

" It appears from the evidence of highly intelligent and " eminent gentlemen of the medical profession, residing in " the neighborhood of the marshes on both sides of the

"Thames below London Bridge, that the diseases preva- "lent in these districts are highly indicative of malarious "influences, fever-and-ague being very prevalent; and "that the sickness and mortality are greatest in those lo- "calities which adjoin imperfectly drained lands, and far "exceed the usual average; and that ague and allied dis- "orders frequently extend to the high grounds in the vicin- "ity. In those districts where a partial drainage has "been effected, a corresponding improvement in the health "of the inhabitants is perceptible."

In the evidence given before the committee, Dr. P. Bossey testified that the malaria from salt marshes varied in intensity, being most active in the morning and in the Summer season. The marshes are sometimes covered by a little fog, usually not more than three feet thick, which is of a very offensive odor, and detrimental to health. Away from the marshes, there is a greater tendency to disease on the side toward which the prevailing winds blow.

Dr. James Stewart testified that the effect of malaria was greatest when very hot weather succeeds heavy rain or floods. He thought that malaria could be carried *up* a slope, but has never been known to descend, and that, consequently, an intervening hill affords sufficient protection against marsh malaria. He had known cases where the edges of a river were healthy and the uplands malarious.

In Santa Maura and Zante, where he had been stationed with the army, he had observed that the edge of a marsh would be comparatively healthy, while the higher places in the vicinity were exceedingly unhealthy. He thought that there were a great many mixed diseases which began like ague and terminated very differently; those diseases would, no doubt, assume a very different form if they were not produced by the marsh air; many diseases are very difficult to treat, from being of a mixed character beginning like marsh fevers and terminating like inflammatory fevers, or diseases of the chest.

Dr. George Farr testified that rheumatism and tic-doloreux were very common among the ladies who live at the Woolwich Arsenal, near the Thames marshes. Some of these cases were quite incurable, until the patients removed to a purer atmosphere.

W. H. Gall, M. D., thought that the extent to which malaria affected the health of London, must of course be very much a theoretical question; "but it is very remark-"able that diseases which are not distinctly miasmatic, do "become much more severe in a miasmatic district. In-"fluenzas, which prevailed in England in 1847, were very "much more fatal in London and the surrounding parts "than they were in the country generally, and influenza "and ague poisons are very nearly allied in their effects. "Marsh miasms are conveyed, no doubt, a considerable "distance. Sufficiently authentic cases are recorded to "show that the influence of marsh miasm extends several "miles." Other physicians testify to the fact, that near the Thames marshes, the prevalent diseases are all of them of an aguish type, intermittent and remittent, and that they are accompanied with much dysentery. Dr. John Manly said that, when he first went to Barking, he found a great deal of ague, but since the draining, in a population of ten thousand, there are not half-a-dozen cases annually and but very little remittent.

The following Extract is taken from the testimony of Sir Culling Eardly, Bart.:

"Chairman:—I believe you reside at Belvidere, in the "parish of Erith?—Yes.—Ch.: Close to these marshes? "—Yes.—Ch.: Can you speak from your own knowledge, "of the state of these marshes, with regard to public "health?—Sir C.: I can speak of some of the results "which have been produced in the neighborhood, from the "condition of the marshes; the neighborhood is in one "continual state of ague. My own house is protected, from "the height of its position, and a gentleman's house is less "liable to the influence of malaria than the houses of the "lower classes. But even in my house we are liable to "ague; and to show the extraordinary manner in which "the ague operates, in the basement story of this house "where my men-servants sleep, we have more than once "had bad ague. In the attics of my house, where my "maid-servants sleep, we have never had it. Persons are "deterred from settling in the neighborhood by the agu-"ish character of the country. Many persons, attracted "by the beauty of the locality, wish to come down and "settle; but when they find the liability to ague, they "are compelled to give up their intention. I may mention "that the village of Erith itself, bears marks of the influ-"ence of malaria. It is more like one of the desolate "towns of Italy, Ferrara, for instance, than a healthy, "happy, English village. I do not know whether it is "known to the committee, that Erith is the village describ-"ed in Dickens' *Household Words*, as Dumble-down-"deary, and that it is a most graphic and correct descrip-"tion of the state of the place, attributable to the unhealthy "character of the locality."

He also stated that the ague is not confined to the marshes, but extends to the high lands near them.

The General Board of Health, of England, at the close of a voluminous report, publish the following "Conclusions "as to the Drainage of Suburban Lands:—

"1. Excess of moisture, even on lands not evidently wet, "is a cause of fogs and damps.

"2. Dampness serves as a medium for the conveyance of "any decomposing matter that may be evolved, and adds "to the injurious effects of such matters in the air:—in "other words the excess of moisture may be said to increase "or aggravate atmospheric impurities.

"3. The evaporation of the surplus moisture lowers the "temperature, produces chills, and creates or aggravates "the sudden and injurious changes or fluctuations by "which health is injured."

In view of the foregoing opinions as to the cause of malaria, and of the evidence as to the effect of draining in removing the unhealthy condition in which those causes originate, it is not too much to say that,—in addition to the capital effect of draining on the productive capacity of the land,—the most beneficial sanitary results may be confidently expected from the extension of the practice, especially in such localities as are now unsafe, or at least undesirable for residence.

In proportion to the completeness and efficiency of the means for the removal of surplus water from the soil:—in proportion, that is, to the degree in which the improved tile drainage described in these pages is adopted,—will be the completeness of the removal of the causes of disease. So far as the drying of malarious lands is concerned, it is only necessary to construct drains in precisely the same manner as for agricultural improvement.

The removal of the waste of houses, and of other filth, will be considered in the next chapter.

CHAPTER XI.

HOUSE DRAINAGE AND TOWN SEWERAGE IN THEIR RELATIONS TO THE PUBLIC HEALTH.

The following is extracted from a report made by the General Board of Health to the British Parliament, concerning the administration of the Public Health Act and the Nuisances Removal and Diseases Prevention Acts from 1848 to 1854.

"Where instances have been favorable for definite ob- "servation, as in broad blocks of buildings, the effects of "sanitary improvement have been already manifested to an "extent greater than could have been anticipated, and than "can be readily credited by those who have not paid atten- "tion to the subject.

"In one favorable instance, that of between 600 and 700 "persons of the working class in the metropolis, during a "period of three years, the average rate of mortality has "been reduced to between 13 and 14 in 1000. In another "instance, for a shorter period, among 500 persons, the "mortality has been reduced as low as even 7 in 1000. "The average rate of mortality for the whole metropolis "being 23 in 1000.

"In another instance, the abolishing of cess-pools and "their replacement by water-closets, together with the "abolishing of brick drains and their replacement by im-

" permeable and self-cleansing stone-ware pipes, has been " attended with an immediate and extraordinary reduction " of mortality. Thus, in Lambeth Square, occupied by a " superior class of operatives, in the receipt of high wages, " the deaths, which in ordinary times were above the gen- " eral average, or more than 30 in 1000, had risen to a rate " of 55 in 1000. By the abolishing of cess-pools, which " were within the houses, and the substitution of water- " closets, and with the introduction of tubular, self-cleansing " house-drains, the mortality has been reduced to 13 in 1000.

" The reduction of the mortality was effected precisely " among the same occupants, without any change in their " habits whatever."

" Sewers are less important than the House-Drains and " Water-Closets, and if not carrying much water, may be- " come cess-pools. In the case of the Square just referred " to, when cess-pools and drains of deposit were removed " without any alteration whatever in the adjacent sewers, " fevers disappeared from house to house, as these recep- " tacles were filled up, and the water-closet apparatus sub- " stituted, merely in consequence of the removal of the de- " composing matter from beneath the houses to a distant " sewer of deposit or open water course.

" If the mortality were at the same rate as in the model " dwellings, or in the improved dwellings in Lambeth " Square, the annual deaths for the whole of the metropolis " would be 25,000 less, and for the whole of England and " Wales 170,000 less than the actual deaths.

" If the reduced rate of mortality in these dwellings " should continue, and there appears to be no reason to " suppose that it will not, the extension to all towns which " have been affected, of the improvements which have been " applied in these buildings, would raise the average age " at death to about forty-eight instead of twenty-nine, the " present average age at death of the inhabitants of towns " in all England and Wales."

The branch of the Art of Drainage which relates to the removal of the fecal and other refuse wastes of the population of towns, is quite different from that which has been described in the preceding pages, as applicable to the agricultural and sanitary improvement of lands under cultivation, and of suburban districts. Still, the fact that town and house drainage affords a means for the preservation of valuable manures, justifies its discussion in an agricultural work, and "draining for health" would stop far short of completeness were no attention paid to the removal of the cause of diseases, which are far more fatal than those that originate in an undrained condition of the soil.

The extent to which these diseases, (of which typhoid fever is a type,) are prevented by sanitary drainage, is strikingly shown in the extract which commences this chapter. Since the experience to which this report refers, it has been found that the most fatal epidemics of the lower portions of London originated in the choked condition of the street sewers, whose general character, as well as the plan of improvement adopted are described in the following "Extracts from the Report of the Metropolitan Board of Works," made in 1866.

"The main sewers discharged their whole contents di-"rect into the Thames, the majority of them capable of "being emptied only at the time of low water; conse-"quently, as the tide rose, the outlets of the sewers were "closed, and the sewage was dammed back, and became "stagnant; the sewage and impure waters were also "constantly flowing from the higher grounds, in some in-"stances during 18 out of the 24 hours, and thus the thick "and heavy substances were deposited, which had to be "afterwards removed by the costly process of hand labor. "During long continued or copious falls of rain, more par-"ticularly when these occurred at the time of high water "in the river, the closed outlets not having sufficient stor-"age capacity to receive the increased volume of sewage, "the houses and premises in the low lying districts, espec-"ially on the south side of the river, became flooded by "the sewage rising through the house drains, and so con-"tinued until the tide had receded sufficiently to afford a "vent for the pent-up waters, when the sewage flowed "and deposited itself along the banks of the river, evolv-"ing gases of a foul and offensive character.

"This state of things had a most injurious effect upon "the condition of the Thames; for not only was the sew-"age carried up the river by the rising tide, at a time "when the volume of pure water was at its minimum, and "quite insufficient to dilute and disinfect it, but it was "brought back again into the heart of the metropolis, there "to mix with each day's fresh supply, until the gradual "progress towards the sea of many day's accumulation "could be plainly discerned; the result being that the por-"tion of the river within the metropolitan district became "scarcely less impure and offensive than the foulest of the "sewers themselves. * * * * * *

"The Board, by the system they have adopted, have "sought to abolish the evils which hitherto existed, by "constructing new lines of sewers, laid in a direction at "right angles to that of the existing sewers, and a little "below their levels, so as to intercept their contents and "convey them to an outfall, on the north side of the Thames "about 11¼ miles, and on the south side about 14 miles, "below London Bridge. By this arrangement as large a "proportion of the sewage as practicable is carried away "by gravitation, and a constant discharge for the remain-"der is provided by means of pumping. At the outlets, "the sewage is delivered into reservoirs situate on the "banks of the Thames, and placed at such levels as enable "them to discharge into the river at or about the time of "high water. The sewage thus becomes not only at "once diluted by the large volume of water in the river at "the time of high water, but is also carried by the ebb

"26 miles below London Bridge, and its return by the fol-"lowing flood-tide within the metropolitan area, is effec-"tually prevented."

The details of this stupendous enterprise are of sufficient interest to justify the introduction here of the "General Statistics of the Works" as reported by the Board.

"A few statistics relative to the works may not prove "uninteresting. The first portion of the works was com-"menced in January 1859, being about five months after "the passing of the Act authorising their execution. "There are 82 miles of main intercepting sewers in London. "In the construction of the works 318,000,000 of bricks, "and 880,000 cubic yards of concrete have been used, "and 3,500,000 cubic yards of earth excavated. The cost, "when completed, will have been about £4,200,000. The "total pumping power employed is 2,300 nominal "horse power: and if the engines were at full work, night "and day, 44,000 tons of coals per annum would be used; "but the average consumption is estimated at 20,000 tons. "The sewage to be intercepted by the works on the north "side of the river, at present amounts to 10,000,000 cubic "feet, and on the south side 4,000,000 cubic feet per day; "but provision is made for an anticipated increase in these "quantities, in addition to the rainfall, amounting to a to-"tal of 63,000,000 cubic feet per day, which is equal to a "lake of 482 acres, three feet deep, or 15 times as large as "the Serpentine in Hyde Park."

A very large portion of the sewage has to be lifted thirty-six feet to the outfall sewer. The works on the north side of the Thames were formally opened, by the Prince of Wales, in April 1865.

In the hope that the immense amount of sewage, for which an escape has been thus provided, might be profitably employed in agriculture, advertisements were inserted in the public journals asking for proposals for carrying out such a scheme; and arrangements were subsequently made for an extension of the works, by private enterprise, by the construction of a culvert nine and a half feet in diameter, and forty miles in length, capable of carrying 12,000,000 cubic feet of sewage per day to the barren sands on the coast of Essex; the intention being to dispose of the liquid to farmers along the line, and to use the surplus for the fertilization of 7000 acres, (to be subsequently increased,) which are to be reclaimed from the sea by embankments and valve sluice-gates.

The estimated cost of this enterprise is about $10,000,000.

The work which has been done, and which is now in contemplation, in England, is suggestive of what might, with advantage, be adopted in the larger cities in America. Especially in New York an improved means of outlet is desirable, and it is doubtful whether the high rate of mortality of that city will be materially reduced before effective measures are devised for removing the vast accumulations of filth, which ebb and flow in many of the larger sewers, with each change of the tide; and which are deposited between the piers along the river-sides.

It would be practicable to construct a main receiving sewer under the river streets, skirting the city, from the vicinity of Bellevue Hospital on the east side, passing near the outer edge of the Battery, and continuing to the high land near 60th street on the west side; having its water level at least twenty feet below the level of the street, and receiving all of the sewage which now flows into the river. At the Battery, this receiving sewer might be connected, by a tunnel, with the Brooklyn shore, its contents being carried to a convenient point south of Fort Hamilton,—where their discharge, (by lifting steam pumps), into the waters of the Lower Bay, would be attended with no inconvenience. The improvement being carried out to this point, it would probably not be long before the advantages to result from the application of the sewage to the sandy soil on the south side of Long Island would be manifest.

The effect of such an improvement on the health of the city,—which is now in constant danger from the putrefying filth of the sewers, (these being little better than covered cess-pools under the streets,)—would, no doubt, equal the improvement that has resulted from similar work in London.

The foregoing relates only to the main outlets for town sewage. The arterial drainage, (the lateral drains of the system,) which receives the waste of the houses and the wash of the streets, is entirely dependent on the outlet sewers, and can be effective only when these are so constructed as to afford a free outfall for the matters that it delivers to them. In many towns, owing to high situation, or to a rapid inclination of surface, the outfall is naturally so good as to require but little attention. In all cases, the manner of constructing the collecting drains is a matter of great importance, and in this work a radical change has been introduced within a few years past.

Formerly, immense conduits of porous brick work, in all cases large enough to be entered to be cleansed, by hand labor, of their accumulated deposits, were considered necessary for the accommodation of the smallest discharge. The consequence of this was, that, especially in sewers carrying but little water, the solid matters contained in the sewage were deposited by the sluggish flow, frequently causing the entire obstruction of the passages. Such drains always required frequent and expensive cleansing by hand, and the decomposition of the filth which they contained produced a most injurious effect on the health of persons living near their connections with the street. The foul liquids with which they were filled, passing through their porous walls, impregnated the earth near them, and sometimes reached to the cellars of adjacent houses, which were in consequence rendered extremely unhealthy. Many such sewers are now in existence, and some such are still being constructed. Not only are they unsatisfactory, they are much more expensive in construction, and require much attention and labor for repairs, and cleansing, than do the stone-ware pipe sewers which are now universally adopted wherever measures are taken to investigate their comparative merits. An example of the difference between the old and modern styles of sewers is found in the drainage of the Westminster School buildings, etc., in London.

The new drainage conveys the house and surface drainage of about two acres on which are fifteen large houses. The whole length of the drain is about three thousand feet, and the entire outlet is through two nine inch pipes. The drainage is perfectly removed, and the pipes are always clean, no foul matters being deposited at any point. This drainage has been adopted as a substitute for an old system of sewerage of which the main was from 4 feet high, by 3 feet 6 inches wide, to 17 feet high and 6 or 7 feet wide. The houses had cess-pools beneath them, which were filled with the accumulations of many years, while the sewers themselves were scarcely less offensive. This condition resulted in a severe epidemic fever of a very fatal character.

An examination instituted to discover the cause of the epidemic resulted in the discovery of the facts set forth above, and there were removed from the drains and cess-pools more than 550 loads of ordure. The evaporating surface of this filth was more than 2000 square yards.

Since the new drainage, not only has there been no recurrence of epidemic fever, but "a greater improvement in "the general health of the population has succeeded than "might be reasonably expected in a small block of houses, "amidst an ill-conditioned district, from which it cannot be "completely isolated."

The principle which justifies the use of pipe sewers is precisely that which has been described in recommending small tiles for agricultural drainage,—*to wit:* that the rapidity of a flow of water, and its power to remove obstacles, is in proportion to its depth as compared with its width. It has been found in practice, that a stream which wends its sluggish way along the bottom of a large brick culvert, when concentrated within the area of a small pipe of regular form, flows much more rapidly, and will carry away even whole bricks, and other substances which were an obstacle to its flow in the larger channel. As an experiment as to the efficacy of small pipes Mr. Hale, the surveyor, who was directed by the General Board of Health of London to make the trial, laid a 12-inch pipe in the bottom of a sewer 5 feet and 6 inches high, and 3 feet and 6 inches wide. The area drained was about 44 acres. He found the velocity of the stream in the pipe to be four and a half times greater than that of the same amount of water in the sewer. The pipe at no time accumulated silt, and the force of the water issuing from the end of the pipe kept the bottom of the sewer perfectly clear for the distance of 12 feet, beyond which point some bricks and stones were deposited, their quantity increasing with the distance from the pipe. He caused sand, pieces of bricks, stones, mud, etc., to be put into the head of the pipe. These were all carried clear through the pipe, but were deposited in the sewer below it.

It has been found by experiment that in a flat bottomed sewer, four feet wide, having a fall of eight inches in one hundred feet, a stream of water one inch depth, runs very sluggishly, while the same water running through a 12-inch pipe, laid on the same inclination, forms a rapid stream, carrying away the heavy silt which was deposited in the broad sewer. As a consequence of this, it has been found, where pipe sewers are used, even on almost imperceptible inclinations, that silt is very rarely deposited, and the waste matters of house and street drainage are carried immediately to the outlet, instead of remaining to ferment and poison the atmosphere of the streets through which they pass. In the rare cases of obstruction which occur, the pipes are very readily cleansed by flushing, at a tithe of the cost of the constant hand-work required in brick sewers.

For the first six or seven hundred feet at the head of a sewer, a six inch pipe will remove all of the house and street drainage, even during a heavy rain fall; and if the inclination is rapid, (say 6 inches to 100 feet,) the acceleration of the flow, caused partly by the constant additions to the water, pipes of this size may be used for considerably greater distances. It has been found by actual trial that it is not necessary to increase the size of the pipe sewer in exact proportion to the amount of drainage that it has to convey, as each addition to the flow, where drainage is admitted from street openings or from houses, accelerates the velocity of the current, pipes discharging even eight times as much when received at intervals along the line as they would take from a full head at the upper end of the sewer.

For a district inhabited by 10,000 persons, a 12-inch pipe would afford a sufficient outlet, unless the amount of road drainage were unusually large, and for the largest sewers, pipes of more than 18 inches diameter are rarely used, these doing the work which, under the old system, was alloted to a sewer 6 feet high and 3 feet broad.

Of course, the connections by which the drainage of roads is admitted to these sewers, must be provided with ample silt-basins, which require frequent cleaning out. In the construction of the sewers, man-holes, built to the surface, are placed at sufficient intervals, and at all points where the course of the sewer changes, so that a light placed at one of these may be seen from the next one;—the contractor being required to lay the sewer so that the light may be thus seen, a straight line both of inclination and direction is secured.

The rules which regulate the laying of land-drains apply with equal force in the making of sewers, that is no part of the pipe should be less perfect, either in material or construction, than that which lies above it; and where the inclination becomes less, in approaching the outlet, silt-basins should be employed, unless the decreased fall is still rapid. The essential point of difference is, that while land drains may be of porous material, and should have open joints for the admission of water, sewer pipes should be of impervious glazed earthen-ware, and their joints should be securely cemented, to prevent the escape of the sewage, which it is their province to remove, not to distribute. Drains from houses, which need not be more than 3 or 4 inches in diameter, should be of the same material, and should discharge with considerable inclination into the pipes, being connected with a curving branch, directing the fluid towards the outlet.

In laying a sewer, it is customary to insert a pipe with a branch opposite each house, or probable site of a house.

It is important that, in towns not supplied with water-works, measures be taken to prevent the admission of too much solid matter in the drainage of houses. Water being the motive power for the removal of the solid parts of the sewage, unless there be a public supply which can be turned on at pleasure, no house should deliver more solid matter than can be carried away by its refuse waters.

The drainage of houses is one of the chief objects of sewerage.

In addition to the cases cited above of the model lodging houses in Lambeth Square, and of the buildings at Westminster, it may be well to refer to a remarkable epidemic which broke out in the Maplewood Young Ladies' Institute in Pittsfield, Mass., in 1864, which was of so violent and fatal a character as to elicit a special examination by a committee of physicians. The family consisted, (pupils, servants, and all,) of one hundred and twelve persons. Of these, fifty-one were attacked with well-defined typhoid fever during a period of less than three weeks. Of this number thirteen died. The following is extracted from the report of the committee:

"Of the 74 resident pupils heard from, 66 are reported "as having had illness of some kind at the close of the "school or soon after. This is a proportion of $\frac{33}{37}$ or nearly "90 per cent. Of the same 74, fifty-one had typhoid fever, "or a proportion of nearly 69 per cent. If all the people "in the town, say 8000, had been affected in an equal pro- "portion, more than 7000 would have been ill during these "few weeks, and about 5500 of them would have had "typhoid fever, and of these over 1375 would have died. "If it would be a more just comparison to take the whole "family at Maplewood into the account, estimating the "number at 112, fifty-six had typhoid fever, or 50 per "cent., and of these fifty-six, sixteen died, or over 28.5 per "cent. These proportions applied to the whole population "of 8000, would give 4000 of typhoid fever in the same "time; and of these 1140 would have died. According "to the testimony of the practising physicians of Pittsfield, "the number of cases of typhoid fever, during this period, "aside from those affected by the influences at Maplewood, "was small, some physicians not having had any, others "had two or three." These cases amounted to but eight, none of which terminated fatally.

The whole secret of this case was proven to have been the retention of the ordure and waste matter from the kitchens and dormitories in privies and vaults, underneath or immediately adjoining the buildings, the odor from these having been offensively perceptible, and under certain atmospheric conditions, having pervaded the whole house.

The committee say "it would be impossible to bring "this report within reasonable limits, were we to discuss the "various questions connected with the origin and propaga- "tion of typhoid fever, although various theoretical views "are held as to whether the poison producing the disease "is generated in the bodies of the sick, and communicated "from them to the well, or whether it is generated in "sources exterior to the bodies of fever patients, yet all "authorities maintain that a peculiar poison is concerned "in its production.

"Those who hold to the doctrine of contagion admit "that, to give such contagion efficacy in the production of "wide spread results, filth or decaying organic matter is "essential; while those who sustain the theory of non- "contagion—the production of the poison from sources "without the bodies of the sick—contend that it has its "entire origin in such filth—in decomposing matter, espe- "cially in fermenting sewage, and decaying human excreta.

"The injurious influence of decomposing azotised matter, "in either predisposing to or exciting severe disease, and "particularly typhoid fever, is universally admitted among "high medical authorities."

The committee were of the opinion "that the disease "at Maplewood essentially originated in the state of the "privies and drainage of the place; the high temperature, "and other peculiar atmospheric conditions developing, in "the organic material thus exposed, a peculiar poison, "which accumulated in sufficient quantity to pervade "the whole premises, and operated a sufficient length of "time to produce disease in young and susceptible per- "sons. * * * * * * To prevent the poison of "typhoid fever when taken into the system, from produc- "ing its legitimate effects, except by natural agencies, "would require as positive a miracle as to restore a severed "head, or arrest the course of the heavenly bodies in their "spheres. * * * The lesson for all, for the future, is "too obvious to need further pointing out; and the com- "mittee cannot doubt that they would hazard little in "predicting that the wisdom obtained by this sad expe- "rience, will be of value in the future management of this "institution, and secure precautions which will forever "prevent the recurrence of such a calamity."

The results of all sanitary investigation indicate clearly the vital necessity for the complete and speedy removal from human habitations of all matters which, by their decomposition, may tend to the production of disease, and early measures should be taken by the authorities of all towns, especially those which are at all compactly built, to secure this removal. The means by which this is to be effected are to be found in such a combination of water-supply and sewerage, as will furnish a constant and copious supply of water to dissolve or hold in suspension the whole of the waste matters, and will provide a channel through which they may be carried away from the vicinity of residences. If means for the application of the sewage water to agricultural lands can be provided, a part if not the whole of the cost of the works will be thus returned.

Concerning the details of house drainage, it would be impossible to say much within the limits of this book. The construction of water-closets, soil-pipes, sinks, etc., are too will be understood to need a special description here.

The principal point, (aside from the use of pipes instead of brick-sewers and brick house-drains,) is what is called in London the system of Back Drainage, where only principal main lines of sewers are laid under the streets, all collecting sewers passing through the centres of the blocks in the rear of the houses. Pipes for water supply are disposed in the same manner, as it is chiefly at the rears of houses that water is required, and that drainage is most necessary; and this adjustment saves the cost, the annoyance and the loss of fall, which accompany the use of pipes running under the entire length of each house. Much tearing up of pavements, expensive ditching in hard road-ways, and interference with traffic is avoided, while very much less ditching and piping is necessary, and repairs are made with very little annoyance to the occupants of houses. The accompanying diagrams, (Figs. 48–49,) illustrate the difference between the old system of drainage with brick sewers under the streets, and brick drains under the houses, and pipe sewers under main streets and through the back yards of premises. A measurement of these two

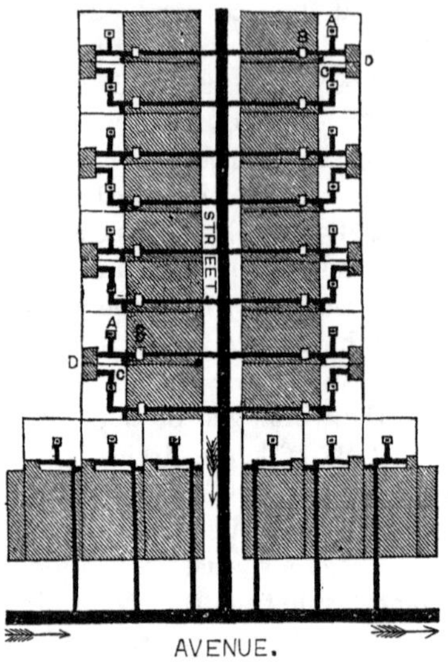

Fig. 48.—OLD STYLE HOUSE DRAINAGE AND SEWERAGE.

methods will show that the lengths of the drains in the new system, are to those of the old, as 1 to $2\frac{1}{4}$;—the fall of the house drains, (these having much less length,) would be 10 times more in the one case than in the other;—the main sewers would have twice the fall, their area would be only $\frac{1}{30}$, and their cubic contents only $\frac{1}{73}$.

Experience in England has shown that if the whole cost of water supply and pipe sewers is, with its interest, divided over a period of thirty years,—so that at the end of that time it should all be repaid,—the annual charge would not be greater than the cost of keeping house-drains and cess-pools clean. The General Board of Health state that "the expense of cleansing the brick house-drains and cess-pools for four or five years, would pay the expense of properly constructed water-closets and pipe-drains, for the greater number of old premises."

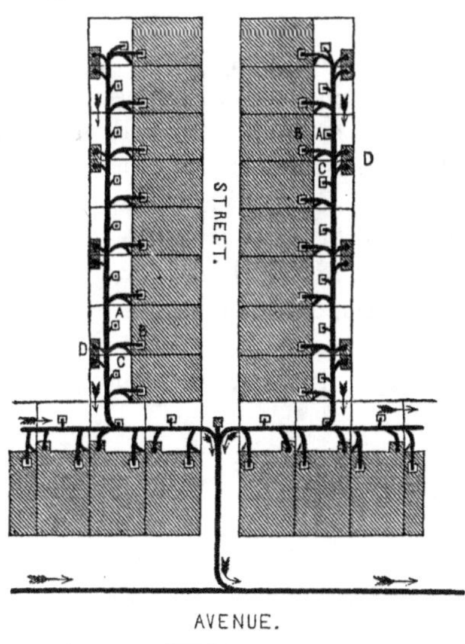

Fig. 49.—MODERN HOUSE DRAINAGE AND SEWERAGE.

One of the reports of this body, which has added more than any other organization to the world's knowledge on these subjects, closes with the following:

"Conclusions obtained as to house drainage, and the sewerage and cleansing of the sites of towns."

"That no population living amidst impurities, arising "from the putrid emanations from cess-pools, drains and "sewers of deposit, can be healthy or free from the attacks "of devastating epidemics.

"That as a primary condition of salubrity, no ordure

"and town refuse can be permitted to remain beneath or "near habitations.

"That by no means can remedial operations be so con- "veniently, economically, inoffensively, and quickly effected "as by the removal of all such refuse dissolved or sus- "pended in water.

"That it has been subsequently proved by the operation "of draining houses with tubular drains, in upwards of "19,000 cases, and by the trial of more than 200 miles of "pipe sewers, that the practice of constructing large brick "or stone sewers for general town drainage, which detain "matters passing into them in suspension in water, which "accumulate deposit, and which are made large enough "for men to enter them, and remove the deposit by hand "labor, without reference to the area to be drained, has "been in ignorance, neglect or perversion of the above "recited principles.

"That while sewers so constructed are productive of "great injury to the public health, by the diffusion into "houses and streets of the noxious products of the decom- "posing matters contained in them, they are wasteful from "the increased expense of their construction and repair, "and from the cost of ineffectual efforts to keep them free "from deposit.

"That the house-drains, made as they have heretofore "been, of absorbent brick or stone, besides detaining sub- "stances in suspension, accumulating foul deposit, and "being so permeable as to permit the escape of the liquid "and gaseous matters, are also false in principle and waste- "ful in the expense of construction, cleansing and repair.

"That it results from the experience developed in these "inquiries, that improved tubular house-drains and sewers "of the proper sizes, inclinations, and material, detain and "accumulate no deposit, emit no offensive smells, and re- "quire no additional supplies of water to keep them clear.

"That the offensive smells proceeding from any works "intended for house or town drainage, indicate the fact "of the detention and decomposition of ordure, and afford "decisive evidence of mal-construction or of ignorant or "defective arrangement.

"That the method of removing refuse in suspension in "water by properly combined works, is much better than "that of collecting it in pits or cess-pools near or under-"neath houses, emptying it by hand labor, and removing "it by carts.

"That it is important for the sake of economy, as well "as for the health of the population, that the practice of "the removal of refuse in suspension in water, and by com-"bined works, should be applied to all houses, especially "those occupied by the poorer classes."

Later investigations of the subject have established two general conclusions applicable to the subject, namely, that:

"*In towns all offensive smells from the decomposition* "*of animal and vegetable matter, indicate the generation* "*and presence of the causes of insalubrity and of prevent-* "*able disease, at the same time that they prove defective* "*local administration;*

"and correlatively, that:

"*In rural districts all continuous offensive smells from* "*animal and vegetable decomposition, indicate prevent-* "*able loss of fertilizing matter, loss of money, and bad* "*husbandry.*"

The principles herein set forth, whether relating to sanitary improvement, to convenience and decency of living, or to the use of waste matters of houses in agricultural improvement, are no less applicable in America than elsewhere; and the more general adoption of improved house drainage and sewerage, and of the use of sewage matters in agriculture, would add to the health and prosperity of its people, and would indicate a great advance in civilization.

INDEX.

DRAINING ENGINEERING.

The undersigned is prepared to assume the personal direction of works of Agricultural and Town Drainage, and Water Supply, in any part of the country; or to send advice and information, by letter, for the guidance of others.

Persons sending maps of their land, with contour lines, (see Fig. 8, page 54,) accompanied by such information as can be given in writing, will be furnished with explicit instructions concerning the arrangement and depth of the drains required; kinds and sizes of tiles to be used; management of the work, etc., etc.

The lines of drains will be laid down, on the maps, for the direction of local engineers,—and, when required, the grades will be calculated and noted at the positions of the stakes.

For particulars, address

GEO. E. WARING, JR.,

P. O. Box 290,

NEWPORT, R. I.

AMERICAN POMOLOGY.

APPLES.

By Doct. JOHN A. WARDER,

PRESIDENT OHIO POMOLOGICAL SOCIETY; VICE-PRESIDENT AMERICAN POMOLOGICAL SOCIETY.

293 ILLUSTRATIONS.

This volume has about 750 pages, the first 375 of which are devoted to the discussion of the general subjects of propagation, nursery culture, selection and planting, cultivation of orchards, care of fruit, insects, and the like; the remainder is occupied with descriptions of apples. With the richness of material at hand, the trouble was to decide what to leave out. It will be found that while the old and standard varieties are not neglected, the new and promising sorts, especially those of the South and West, have prominence. A list of selections for different localities by eminent orchardists is a valuable portion of the volume, while the Analytical Index or *Catalogue Raisonné*, as the French would say, is the most extended American fruit list ever published, and gives evidence of a fearful amount of labor.

CONTENTS.

Sent Post-Paid. Price $3.00.

ORANGE JUDD & CO., 41 PARK ROW, NEW-YORK.

www.ingramcontent.com/pod-product-compliance
Lightning Source LLC
LaVergne TN
LVHW050512100826
845148LV00002B/308

9781425576936